T3-AJJ-711

WITHDRAWN

The Coming Robot Revolution

Yoseph Bar-Cohen • David Hanson
Adi Marom, Graphic Artist

The Coming Robot Revolution

Expectations and Fears About Emerging Intelligent, Humanlike Machines

 Springer

Yoseph Bar-Cohen
Jet Propulsion Laboratory (JPL)
California Institute of Technology
Pasadena, CA
yosi@jpl.nasa.gov

David Hanson
Hanson Robotics
Richardson, TX
david@hansonrobotics.com

ISBN 978-0-387-85348-2   e-ISBN 978-0-387-85349-9
DOI 10.1007/978-0-387-85349-9

Library of Congress Control Number: 2008942430

© Springer Science+Business Media, LLC 2009
All rights reserved. This work may not be translated or copied in whole or in part without the written permission of the publisher (Springer Science+Business Media, LLC, 233 Spring Street, New York, NY 10013, USA), except for brief excerpts in connection with reviews or scholarly analysis. Use in connection with any form of information storage and retrieval, electronic adaptation, computer software, or by similar or dissimilar methodology now known or hereafter developed is forbidden.
The use in this publication of trade names, trademarks, service marks, and similar terms, even if they are not identified as such, is not to be taken as an expression of opinion as to whether or not they are subject to proprietary rights.

Printed on acid-free paper

springer.com

Preface

Making a robot that looks and behaves like a human being has been the subject of many popular science fiction movies and books. Although the development of such a robot faces many challenges, the making of a virtual human has long been potentially possible. With recent advances in various key technologies related to hardware and software, the making of humanlike robots is increasingly becoming an engineering reality.

Development of the required hardware that can perform humanlike functions in a lifelike manner has benefitted greatly from development in such technologies as biologically inspired materials, artificial intelligence, artificial vision, and many others. Producing a humanlike robot that makes body and facial expressions, communicates verbally using extensive vocabulary, and interprets speech with high accuracy is extremely complicated to engineer. Advances in voice recognition and speech synthesis are increasingly improving communication capabilities. In our daily life we encounter such innovations when we call the telephone operators of most companies today.

As robotics technology continues to improve we are approaching the point where, on seeing such a robot, we will respond with "Wow, this robot looks unbelievably real!" just like the reaction to an artificial flower. The accelerating pace of advances in related fields suggests that the emergence of humanlike robots that become part of our daily life seems to be imminent. These robots are expected to raise ethical concerns and may also raise many complex questions related to their interaction with humans.

This book covers the reality and the vision in the development and engineering of humanlike robots. The topic is described from various angles, including the state of the art, how these robots are made, their current and potential applications, and the challenges to the developers and users, as well as the concerns and ethical issues. This book includes discussion of the state-of-the-art trends, challenges, benefits, and plans for future developments. In the opening chapter, a distinction is made between humanoid robots that have the general appearance of humans and humanlike robots with an appearance that is identical to humans. Chapter 2 describes the currently available crop of humanoids and humanlike robots, while Chapter 3 examines various components that are involved in making such robots. The subjects of prosthetics, exoskeletons, and bipedal ambulators are covered in Chapter 4. Exoskeleton structures are used to augment the ability of humans in walking. Further, ambulators are chairs with two legs that carry humans and that are able to walk. They were developed to replace wheelchairs for operation in certain difficult-to-maneuver areas, including climbing stairs while carrying a human.

Chapter 5 considers the issues that result from our making robots that mirror humans so closely. These robots challenge our human identity and our primacy as the lead species on the planet. Besides becoming household appliances these robots may significantly impact our lives and our economy; the potential impacts are discussed in Chapter 6. Once such robots become intelligent and perhaps even conscious, we will have to deal with certain ethical issues and others concerns that are expected to arise as described in Chapter 7. The book concludes with a chapter that describes and discusses the capabilities and challenges in developing the technology of humanlike robots.

Yoseph Bar-Cohen, JPL
Pasadena, CA

Acknowledgements

The authors would like to thank those who contributed to the preparation of this book, including the individuals who helped advance the technology that is reported. In particular, the authors would like to thank Gabor Kovacs of Swiss Federal Laboratories, for Materials Testing and Research, and EMPA in Dubendorf, Switzerland, for helping to obtain photos of the humanlike mechanical "Writer" created by Pierre Jaquet-Droz, 1774. Thanks also to Federico Carpi, University of Pisa, Italy, for helping to obtain images of the Leonardo da Vinci's robot drawings. Paul Averill, at JPL, helped to identify sources of information about the da Vinci's robot design. The authors would like to thank Dan Ferber, of *Popular Science*, for his help in the initial stages of planning this book. Jinsong Leng of China helped to obtain the photo of roboticist Zou Renti and his clone robot. We are grateful to Heather Heerema, of Hanson Robotics for her grammatical editing of some of the chapters of this book. Giorgio Metta, from the University of Genova, Italy, provided information about the European iCub robot. Geoff Spinks, the University of Wollongong, Australia, helped to locate an expert in robotic ethics for the peer review of Chapter 7. Also, the principal author would like to thank his wife, Yardena Bar-Cohen, for her useful suggestions and for taking some of the photos that were used to create several of the figures in this book.

Some of the research reported in this book was conducted at the Jet Propulsion Laboratory (JPL), California Institute of Technology, under a contract with the National Aeronautics and Space Administration (NASA).

The authors would also like to acknowledge and express their deepest appreciation to the following individuals, who took the time to review various chapters of this book. These individuals contributed significantly with their comments, constructive criticisms, and recommendations, all of which were very helpful in adding to the value of this book.

David Bruemmer, Idaho National Lab, Idaho Falls, ID
Susan Dodds, University of Wollongong, Australia
Federico Carpi, University of Pisa, Italy
Chad (Odest Chadwicke) Jenkins, Brown University, Providence, RI
Brett Kennedy, Jet Propulsion Laboratory, Pasadena, CA
Kwang Kim, University of Nevada-Reno, Reno, NV
David Kindlon, McCarthy Studios, Baldwin Park, CA

Richard Landon, Stan Winston Studio, Van Nuys CA
Zhiwei Luo, University of Kobe, Japan
Roger Mandel, Rhode Island School of Design, Providence, RI
Nikolaos Mavroidis, United Arab Emirates University, Al-Ain, United Arab Emirates
Chris Melhuish, University of the West of England, Bristol, UK
Peter Plantec, Columnist, VFXworld.com, Hollywood, CA
Joseph Rosen, Dartmouth-Hitchcock Medical Center, Lebanon, NH
Martine Rothblatt, United Therapeutics Corp., Silver Spring, MD
Rick (Richard) Satava, University of Washington Medical Center, Seattle, WA
Scaz (Brian) Scasselati, Yale University, New Haven, CT
Gianmarco Veruggio, Scuola di Robotica, Genova, Italy
Chris Willis, Android World Inc., Denton, TX

The photo, which is showing a robotic head and hand on the back cover of this book, was taken at JPL. The head was made by the coauthor, David Hanson, and the hand was provided to the principal author, Yoseph Bar-Cohen, as a courtesy of Graham Whiteley, Sheffield Hallam University, UK.

Contents

About the Authors

Dr. Yoseph Bar-Cohen is a Senior Scientist and Group Supervisor at the Jet Propulsion Lab, NASA/Caltech, specializing in electroactive materials and devices as well as biomimetic mechanisms. Known best for his pivotal role in the development of artificial muscles, in 2003 *Business Week* entitled him as one of five technology gurus who are "pushing tech's boundaries." Dr. Bar-Cohen received his Ph.D. in physics (1979) from Israel's Hebrew University in Jerusalem. Some of his notable discoveries include the leaky Lamb waves (LLW) and the polar backscattering (PBS) phenomena in composite materials. He has (co)authored over 300 publications, made numerous presentations at national and international conferences, (co)chaired 37 conferences, has 19 registered patents, and is the (co)editor of 4 books. He was named a Fellow of the American Society for Nondestructive Testing (ASNT) in 1996 and of The International Society for Optical Engineering (SPIE) in 2002. Also, he is the recipient of two NASA Honor Award Medals – NASA Exceptional Engineering Achievement Medal (2001) and NASA Exceptional Technology Achievement (2006), plus two SPIE's Lifetime Achievement Awards as well as many other honors and awards.

Dr. David Hanson is an artist/scientist who creates realistic humanoid robots (a.k.a. androids), which are noted for being conversationally intelligent, energy efficient, and designed as novel works of character art/animation. In 2005, the low-power mobility of Hanson's robots was demonstrated in the world's first expressive walking humanoid, an Einstein portrait called "Albert Hubo," appearing on the cover of Wired magazine in January 2006. In addition to hardware innovations, Hanson and his company (Hanson Robotics Inc.) are known for developing increasingly intelligent conversational personas, integrating many forms of artificial intelligence (AI), including speech recognition software, natural language processing, computer vision, and Hanson's own AI systems to hold naturalistic conversations. Hanson has received awards in both art and engineering, including the Cooper Hewwit Triennial award, the National Science Foundation STTR award, and a TX Emerging Technologies Award. Hanson received a BFA from the Rhode Island School of Design in 1996, and his Ph.D. from the University of Texas at Dallas in 2007.

Adi Marom is a designer/artist with a broad international education and work experience. She specializes in the design of interactive kinetic applications. Her work integrates biomimetic technology, applying natural mechanism into deployable designs.

She holds a Masters of Design Engineering from The University of Tokyo, Japan; and a B.A. in Design from Bezalel Academy of Arts and Design, Israel. Currently, she is a scholar at NYU's Interactive Telecommunication Program (ITP). Marom experience consists of working for prominent design studios in Israel, Japan, and the United States. Her artwork has been displayed in exhibitions worldwide and has been featured in inernational media publications, including BoingBoing.net, TrendHunter.com, InventorSpot.com, Casa Brutus (Japan), DAMn° Magazine (Belgium), Joong Ang Daily (Korea), and TimeOut (Israel). This book is Marom's second collaboration with Dr. Yoseph Bar-Cohen. Previously, her designs have been featured in his book "Biomimetics: Biologically-Inspired Technologies", which was published by CRC Press in November 2005.

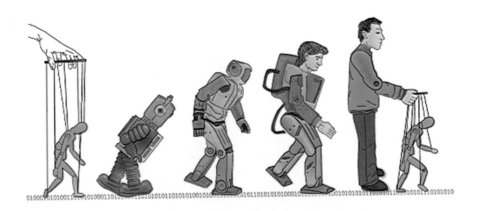

010001010100110101010001101010110101011010101001010101010101101101010101000101010110101010101000101010101010110110101010101010110101010101010

Chapter 1
Introduction

Imagine you are having a polite conversation with a receptionist when you check into a hotel where you suddenly get the feeling that something is weird. In a flash you realize what's wrong – this is not a real person but rather a robot. Your first reaction would probably be "It's unbelievable – she looks so real," just as you would react to an artificial flower that is a good imitation. With a flower, though, you can touch it to find out if it is real; here, you must rely on your other senses to confirm your suspicion.

This science fiction scenario is rapidly approaching reality, as the trend in the development of humanlike robots continues. An illustration of a humanlike robot is given in Figure 1.1, where externally the robot looks like human. Although this figure shows a rendered image of a human and a simulated internal hardware, the humanlike robots today are being made to look relatively close to lifelike.

Since the Stone Age, people have used art and technology to reproduce the human appearance, capabilities, and intelligence. Realistic humanlike robots and simulations, which once seemed just a fantastic, unattainable extension of these efforts, are starting literally to walk into our lives, thanks to recent advances in the development of related technology. Such robots originate from the efforts to adapt and imitate, inspired by nature or more specifically using biology as a model for mimicking. A related field known as "biomimetics" involves the study and the engineering of machines that display the appearance, behavior, and functions of biological systems.

Robots that have humanlike features have been given many names, including humanoids, androids, and automatons. There are many other terms that are used to describe humanlike robots, but the following definitions show the basic distinctions

Y. Bar-Cohen, D. Hanson, *The Coming Robot Revolution*, DOI 10.1007/978-0-387-85349-9_1,
© Springer Science+Business Media, LLC 2009

Figure 1.1. An illustration of a humanlike robot and its "internal organs." Robots are increasingly being made to look lifelike and operate like humans. The human face is a photo of the graphic artist Adi Marom.

between humanoids and humanlike robots, while Table 1.1 lists the wide variety of names and terms that identify various robotic machines with human features.

HUMANOIDS

Robots that have a somewhat human appearance, with a general shape that includes a head, hands, legs, and possibly eyes, are called humanoids. These are fanciful and easily identified machines that are obviously robots (e.g., making them look like astronauts with a helmet-shaped head). The task of roboticists who are making such robots is relatively easy, and it involves fewer requirements than dealing with the complex issues associated with making completely humanlike machines. Such robots include the robot head, Kismet, by Cynthia Breazeal (see Figure 1.2) and the Female Type robot

Table 1.1. Widely used terms that identify various robotic machines with human features.

Term	Description
Android or Zombie	Science fiction creature, mostly a robot that looks like human male
Anthropomorphic machine	A machine that has the attributes of human characteristics. The word was derived from the Greek words *anthropos,* which means human, and *morph,* which means shape or form
Automaton	Mechanical human
Bionic human or Cyborg	A human with a mixture of organic and mechanical components
Gynoid, Fembot, and Feminoid	A robot that looks like human female
Human assistive devices	Prosthetics, exoskeletons, and walking chairs using two legs
Humanlike robot	Synthetic human, artificial human, or robots that look very similar to humans
Humanoid	Intelligent mechanical human. A robot with general human features including a head, a torso, hands, and legs, but has no detailed facial features

(see Figure 1.3), which was made by Tomotaka Takahashi, Robo-Garage, in Kyoto, Japan. Kismet clearly looks like a machine with animal-like ears, but it is included in this chapter since the expressions it makes are very humanlike. It is interesting to note that the Kismet's facial expressions were designed to represent correct social behavior, and that these expressions are generated by computer models of cognition that allow artificially simulating a human's perception, attention, emotion, motivation, behavior, and expressive movement.

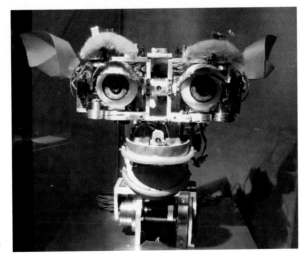

Figure 1.2. The autonomous robot head, Kismet, was developed by Cynthia Breazeal at the MIT Artificial Intelligence Lab. Photo courtesy of Sloan Kulper, Boston, MA, who photographed this robot at the MIT Museum http://web.mit.edu/sloan2/kismet/

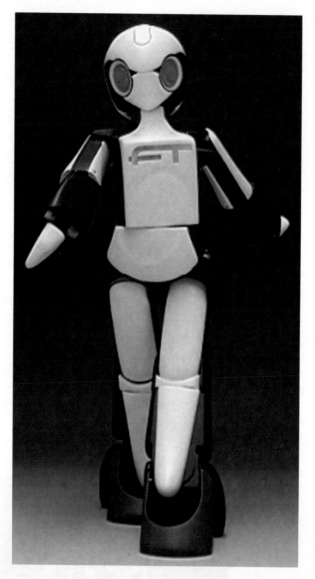

Figure 1.3. The Female Type by RoboGarage is an example of a robot that can perform functions emulating humans. Photo courtesy of Tomotaka Takahashi, Robo-Garage, Kyoto, Japan.

HUMANLIKE ROBOTS

These are machines that are barely distinguishable from real humans; here, roboticists are making every effort to copy the appearance and behavior of humans as realistically as possible. Roboticists building these kinds of robots are mostly from Japan, Korea,

Figure 1.4. The humanlike female robot Repliee Q2 that was developed by Hiroshi Ishiguro of Osaka University and Korkoro Co., Ltd. Photo courtesy of Hiroshi Ishiguro, Osaka Univ. (Ishiguro Lab) and Korkoro Co., Ltd., Japan.

and China, with few found in the United States. Examples of developed humanlike robots are seen in Figures 1.4 and 1.5, and these show how closely the human appearance has been copied. The female humanlike robot that is shown in Figure 1.4 is such a close imitation that it is not easy to determine from the photo if it is a machine or a real human. In Figure 1.5 the roboticist Hiroshi Ishiguro, from Japan, and his replica in the humanlike self-image robot called Geminoid are shown, and the similarity in their appearance is quite impressive

As humanlike robots become more capable and useful, one can envision that years from now they may become our household appliances or even our peers, and we may use them to perform difficult and complex tasks as well as possibly to replace unskilled

Figure 1.5. The roboticist Hiroshi Ishiguro replicated in the humanlike self-image robot called Geminoid. Which one is a real person cannot be easily distinguished. Photo courtesy of Hiroshi Ishiguro, ATR Intelligent Robots and Communication Laboratories, Japan.

human laborers. However, truly humanlike machines may raise fear and dislike, as predicted in 1970 by the Japanese roboticist Masahiro Mori (see also discussions on this in Chapters 5 and 7). In his hypothesis, which is known as the Uncanny Valley, Mori suggested that as the degree of similarity between robots and humans increases, there will be great excitement at first with the progress being made, but when this similarity becomes quite close it will turn to a strong rejection and dislike (graphically described as a valley in the attitude). Eventually, once the similarity reaches an even higher level, then the attitude towards these robots will once again turn to liking.

Besides the issue of attitude towards these robots, with the increase in similarity ethical questions and concerns will raise many unanswered questions: Will these machines complicate our lives or possibly hurt us? Will humanlike robots equipped with artificial cognition continue to be helpful to us or turn against us? Also, we may want to consider how realistic we can make these robots, how realistic we want them to be, and what are the potential dangers of making robots that are humanlike. We need to look at other questions as well, including how much we want to allow such robots to influence our lives, and how can we prevent accidents, deliberate harm, or their use in committing crimes. With regard to the latter, one may wonder what would happen if robots take on certain unique roles, such as serving as clones for specific humans, or have access to our assets and private/intimate information, which they could possibly release to the public or to individuals whose intentions are questionable.

BRIEF HISTORICAL PERSPECTIVE

The word "robot" refers to an electromechanical machine that has biomimetic components and movement characteristics, which give it the ability to manipulate objects and sense its environment along with a certain degree of intelligence. This word was first used in 1921 by the Czech writer Karel Čapek in his play *Rossum's Universal Robots* (R.U.R.). The word "robot" comes from the Czech (also Slovak) word *robota,* which means "hard work" or "slavery." The meaning of the word has evolved and became increasingly associated with intelligent mechanisms that have a biologically inspired shape and functions. Nowadays, the word suggests to a great degree a machine with humanlike characteristics.

Even though marionettes are not formally considered robots (see Figure 1.6), it is worthwhile to mention them because they may be viewed as a precursor to modern humanlike robots. Marionettes are puppets that originated in France in medieval times and were then adapted for use in box, curtain, and black light theaters. Pinocchio is one of the most famous among the many marionettes that were used in shows. Efforts to mechanize the operation of marionettes were attempted by various artists and engineers. In the 1960s, TV producer Gerry Anderson and his colleagues pioneered a technique called "supermarionation," which combined marionettes with electronic components. The developed marionettes performed complex movements, including the making of facial expressions.

The idea of making humanlike machines can be traced as far back as the ancient Greeks, where the god of metalsmiths, Hephaistus, created his own mechanical helpers. These helpers assumed the form of living young women that were strong, vocal, and intelligent (Rosheim, 1994). In the sixteenth century the Jewish legend of the Golem was introduced. In this legend, Rabbi Judah Loew, who is also known by the nickname the *Maharal* of Prague, brought to life a humanlike servant made of clay. Another famous humanlike fictional character is the monster from Mary Shelley's novel *Frankenstein* (1818). In this novel, a scientist named Victor Frankenstein assembles a monster from body parts and brings him to life. Both the Golem and Frankenstein's monster were made to look like living humans and ended up becoming violent, with disastrous consequences. The behavior of these humanlike creations suggests the potential for evil that could result if humanlike forms are given freedom to act without

Figure 1.6. An example of a marionette mounted on a set of strings and is being manipulated to perform in a street show. Photo by Yoseph Bar-Cohen, modified by graphics artist Adi Marom.

restraints and sufficient controls. The concerns and ethical issues related to the development of humanlike robots are considered and discussed in Chapter 7 of this book.

Leonardo da Vinci is credited as being the first person to make a sketch or plan for producing a humanlike machine. In about 1495, da Vinci used his knowledge of the human body to design a mechanical knight that could sit, wave its arms, and move its head via a flexible neck while opening and closing its jaw. This mechanical device is also called Leonardo's robot. It became widely known after an attempt that was made in the 1950s to physically produce the robot using the original sketch (see Figure 1.7).

The first physical machines to appear and act like humans were made in the eighteenth century. In 1737, the French engineer and inventor named Jacques de Vaucanson produced the first such humanlike machine, which was a complete automaton. This life-size mechanical figure played a flute and was called "The Flute Player." This machine was made to play a repertoire of 12 pieces, including "Le Rossignol" (The Nightingale) by Blavet. After few modifications to his machine, de Vaucanson presented it in 1738 to the French Academy of Sciences. In the later part of the same year, he produced another humanlike machine that is called "The Tambourine Player."

Another one of the early machines designed to appear and act like humans was the mechanical "Writer" that was completed in 1772. This machine can be seen at the Musée

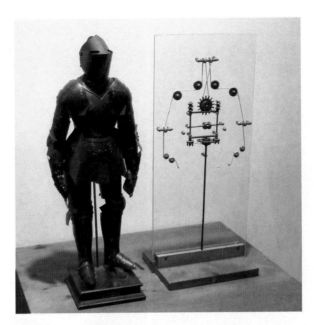

Figure 1.7. Model of Leonardo's robot and its inner working components. Photo is a public domain graphics courtesy of Wikipedia http.//en.wikipedia.org/wiki/Image:Leonardo-Robot3.jpg

d'Art et d'Histoire of Neuchâtel in Switzerland. The "Writer" is 71 cm (28 inches) high and carved of wood by the Swiss clockmaker Jacquet-Droz (see Figure 1.8). This humanlike machine emulates a young boy writing at his desk. When the mechanism starts, the boy dips a feather into an inkwell, shakes the feather twice, places his hand at the top of a page, and writes. The eyes of the "Writer" follow the text while he writes, and when taking ink his head moves to look at the process. The "Writer" is able to write custom text up to 40 letters in length that are coded on wheels with a variety of shapes and sizes.

Designers of automatic mechanisms in the early part of the twentieth century are credited with some of the pioneering efforts in the development of modern intelligent humanlike robots. By mechanizing redundant tasks in the form of automated production lines, engineers improved the manufacturing speed, quality, and uniformity of products and reduced the cost per unit. Robotic mechanisms emerged from developers' efforts to adapt more quickly to changing requirements. Industry was slow, however, to adopt robotic systems (such as manipulator arms), since they were too bulky and expensive, and it required a significant effort to implement, maintain, modify, and/or upgrade them. However, significant advances in the technology and the cost benefits of industrial robots eventually outweighed the complications, and today such robots are standard manufacturing tools in many industries, including the automotive and pharmaceutical industries.

The greatest impact on the use of robots in industry came from improvements in their software and computer controls. Progress in developing powerful microprocessors with high computational speed, very large memory, wide communications bandwidth, and more intuitive and effective software profoundly changed the development of intelligent

Figure 1.8. The humanlike mechanical "Writer" created by Pierre Jaquet-Droz, 1774. The close-up on the right is showing the eyes of the "Writer" follow the writing action. Photo courtesy of the Musée d'Art et d'Histoire, Neuchâtel, Switzerland. (Photo by Mahn Stfano Lori).

robots. Rapid processing of instructions and effective control methodologies allowed for the development of increasingly sophisticated robots with biologically inspired appearance and performance. An example of a robotic arm grabbing onto an object (a DVD packet) is shown in Figure 1.9, which illustrates the biologically inspired nature of this type of manipulator.

Figure 1.9. A robotic arm (made by Barrett Technology Inc.) is shown holding an object illustrating the biologically inspired operation of such manipulators. Photo by Yoseph Bar-Cohen at the 2008 IEEE International Conference on Robotics and Automation (ICRA) held in Pasadena, California.

Besides working on the electromechanical aspects of the full body of humanlike robots, scientists and engineers sought to develop various features to allow for greater mimicking of human functions. These features included speech, vision, sensing, and intelligence. One of the earliest mechanisms of synthesized speech was developed in the mid-nineteenth century by Joseph Faber, who spent 17 years making and perfecting his machine. He called his humanlike mechanical device Euphonia and presented it to the public for the first time on December 22, 1845, at the Philadelphia Musical Fund Hall. Large bellowers behind the humanlike head of his machine were used to blow air into a series of mechanisms that imitated the human lungs, larynx, and mouth. Faber operated his figure using foot pedals and a keyboard, and, in a monotonous voice the figure uttered phrases and even sentences.

Advancements in computers and microprocessors played a very important role in the development of the robot brain and artificial intelligence (AI), which led to smart robots. The Analytical Engine is considered the first machine to have performed complex computational tasks. It was developed by Charles Babbage and Ada Byron in the latter part of the nineteenth century. Even though it was never completed, this device is considered the mechanical predecessor of modern digital computers. The era of digital computers began with the ENIAC computer in 1946, which was the first large-scale general-purpose electronic digital computer. The first time the possibility of building machines that could think and learn was raised in a paper entitled "Computing Machinery and Intelligence" by Alan Turing in 1950. That same year, Grey Walter for the first time showed that there could be navigation and interaction among computerized robotic mechanisms. Walter used two robotic tortoises that moved towards a light source and communicated with each other.

The complexity and unpredictability of natural environments make it virtually impossible to explicitly preprogram a robot for every foreseeable circumstance, and therefore the robot needs to be able to deal with complex situations on its own as well as adapt and learn from its own experience. In order to develop such sophisticated capabilities, researchers in the fields of AI and robotics drew on models, concepts, and methodologies that were inspired and guided by nature, by living creatures that have impressive agility, robustness, flexibility, efficiency, and the ability to learn and adapt in order to survive in the real world.

The increasing introduction of tools for autonomous operation, as well as more humanlike materials, morphology, movement, and functionality are helping to make today's robots more realistic. Even the related software increasingly resembles the organization and functionality of the human central nervous system, and it helps robots perceive, interpret, respond, and adapt to their environment more like humans. An example of a biologically inspired technique is the genetic algorithm, which emulates the survival of the fittest in nature. This algorithm is a widely used computer search technique for finding exact or approximate solutions to optimization problems.

Since the beginning of the era of microprocessors and computers numerous robotic inventions have been conceived, developed, and demonstrated. These ongoing efforts led to increasingly smarter robots that are entering people's lives as products in the areas of recreation and entertainment, education, healthcare, domestic robotic assistance, military, and many others. With the rise in the number of applications, the market is displaying ever more exciting and new humanlike robots. The entertainment industry in particular has sparked a growing number of commercially available humanlike robotic toys that are really lifelike. Toy makers have begun to collaborate with scientists to make characters in movies appear more realistic and to move more like people. Researchers in the field of robotics are also increasingly collaborating with artists to make their robots appear more expressive, believable, and warm. As we interact with robots and make them more sophisticated and humanlike, we increasingly improve our understanding of ourselves – scientifically, socially, and ethically.

An important niche for the application of robots can be the performance of dangerous tasks in hazardous environments, from planetary or deep ocean exploration to operating in areas with toxic gases, radioactivity, dangerous chemicals, bad odors, biological hazards, or extreme temperature environments. Robots can be used to clean up hazardous waste, sweep mine fields, remove bombs, and perform risky search and rescue operations. Conditions such as these require autonomous operation and a robot able to perceive its environment, make decisions, and independently perform tasks.

THE STATE-OF-THE-ART: FICTION AND REALITY

Science fiction books and movies depict humanlike robots that are far beyond today's capability. However, the ideas that are described may be used to guide future development of this technology and also to alert us to the negative possibilities and dangers. With the increases in our ability to make humanlike robots more lifelike, there is a growing concern that they will be used for improper tasks. Ethical and philosophical issues and challenges are being raised and efforts are being made to address these issues before they materialize. In his famous three laws of robotics, to which he later added his Zero Law, the well-known science fiction writer, Isaac Asimov, provided guidelines for human–robot relations. In these proposed "laws," he suggested that robots will be in the role of servants and should not be allowed to cause harm or injury to humans.

The latest advances in technology have led to robots that look very much like humans but are mostly able to perform only limited functions. Some of the recent advances allow them to improve themselves even after they are produced. In other words, they can self-learn and obtain periodic updates. The sophistication of these robots includes fully autonomous operation and self-diagnostics. In the future, they may even be designed to go on their own to a selected maintenance facility for periodic checkups and to be repaired as needed. In case of damage, future robots may be made of biomimetic materials that are capable of self-healing.

Progress in voice synthesis, detection, and recognition for interaction between humanlike robots and humans is enabling robots to communicate verbally, use facial expressions, express emotions while making eye contact, and respond to emotional and verbal cues. These capabilities are making such robots appear and behave in a more natural manner, as they interact and communicate with people in ways that are familiar to humans without the need for training. Some of the human behaviors that robots are capable of performing include nodding when listening to someone speaking to them and periodically having the eyes blink as well as look at the speaker in the eye for brief exchanges. As with human conversation cues, these robots are designed to appear to be listening while not staring into the speaker's eyes for extended periods to avoid making the communicating person uncomfortable.

Today, the conversations that can take place between humans and robots are limited in vocabulary and content. However, much research is dedicated to developing machines that can understand human conversation. This has many applications, including converting voice to computer text, as well as translating foreign languages and dialects. As development progresses, one may envision the possibility of humanlike robots discussing strategies related to social issues, business investment, and personal problems, or possibly even hold philosophical and political debates.

As household appliances, or "companions," they must be able to support us in our daily tasks, which range from cleaning the house, repairing and performing routine maintenance of household appliances, as well as possibly safeguarding the house and its perimeter. These tasks may sound menial, but they are actually quite complex and require greater cognitive capabilities than today's robots are not to offer. In the role of guards, robots may be able to check visitors to see if they are welcome relatives, friends, or expected service personnel. They may also have the ability to identify unwelcome individuals and, if necessary, notify law enforcement agents and keep suspects from leaving until the police arrive. A robot may be able to cheer you up, laugh when a funny situation occurs, smell and identify odors, as well as taste food, suggest recipes, and provide detailed nutrition and health information. Also, such robots may verbally provide the latest news, alert you to important e-mails, or provide other interesting information expressed in any desired language, accent, or gender voice and translate or use sign language if needed.

Some household robots have already emerged commercially and are used widely in the United States and Japan. But these machines do not have a humanlike form yet. An example of such robots includes the disk-like self-guided Roomba^{TM} (made by iRobot Corp.) for vacuum cleaning, with several million units sold so far. Another example is the lawnmower called the Robomow^{TM} that mows the lawn while avoiding obstacles.

The entertainment industry has already benefited from this trend in realistic lifelike robots, with many humanlike toys and special effects in movies. An example of such a toy includes the Mattel's 2001 Miracle Moves Baby, which was developed in partnership with TOYinnovation, Inc. This doll wakes up, yawns, appears tired, sucks her bottle or her thumb, giggles, burps, and can be rocked to sleep in a lifelike way. Another humanlike doll is Hasbro's Baby Alive doll that has been on the market since 2006 (see Figure 1.10).

In addition to robots being created for entertainment, support staff in the healthcare industry is benefiting from this technology. A nursemaid robot and others were developed in Japan and the United States to assist recovering patients, the elderly, and other people who need physical or emotional support. Further details about the currently available robots are described in Chapter 2; potential development and applications are discussed in Chapter 6.

Robots that are remotely controlled by a human operator are being developed for potential future space missions. A robotic astronaut, called Robonaut, was designed by

Figure 1.10. The Hasbro's 2006 Baby Alive is an expressive doll. Photo by David Hanson, Hanson Robotics LLC.

the Robot Systems Technology Branch at NASA and Johnson Space Center in collaboration with Defense Advanced Research Projects Agency. This robot (see Figure 1.11) can help humans work and explore space, as well as perform extravehicular activity. The design of Robonaut eliminates the need for specialized robotic tools to perform in-orbit robotics. It takes advantage of human abilities by keeping the human operator in the control loop in the form of tele-presence where the human controls the operation via mirroring his or her action at the remote site by the Robonaut.

The importance of robotics as the next hot technology was highlighted in the January 2007 cover article that was published in *Scientific American*. This article was authored by Bill Gates, the chairman of Microsoft Corporation, who is considered one of the leaders of the personal computer revolution. In this article Gates expressed great confidence that robots would become part of every home. He pointed out that robotic hardware and software are still incompatible, in a way, similar to how computers and software struggled when he started his company. He suggested that in order to achieve widespread use of these machines, robot developers need standard hardware and software tools so that they do not have to "start from scratch" each time a new robot is developed.

Figure 1.11. The humanlike robot called Robonaut is capable of performing manipulation tasks via human tele-presence control. Photo courtesy of NASA Johnson Space Center, Houston, TX.

Even though humanlike robots are increasingly equipped with significantly improved capabilities, there are still many issues that limit their widespread use, including the very high cost of the sophisticated ones, their relatively short battery charge, and their limited functionality. Once such robots reach mass production levels, their price is expected to come down to more affordable levels. When such robots become less expensive, more useful, and safer to operate, they will probably become more attractive as household helpers for performing human-related services.

WHY MAKE HUMANLIKE ROBOTS?

Humans have built the world around them in a way that it is ideal for use by their body size, shape, and abilities, and everything in daily life is adapted in this way to the human form. Examples include our homes, workplaces, and facilities, the tools we use, our means of transportation, our instruments, the height at which we keep our stuff, and so on. Thus, if we want to develop robots that would best support us, it would be better to make them as much as possible a replica of our shape, average size, and ability. Thus, these robots could reach the handle to open doors, look at humans at eye level, climb stairs, sit on our chairs, use our desks to perform various tasks (including writing or reading if needed), repair and perform maintenance of our home appliances, enter our automobile and sit in it or even possibly drive it, open our closet to bring us requested objects, bring us books from our bookshelves, and perform many other support tasks as expressed verbally or via a remote control device.

Humanlike Form Allows Us to Better Understand Robots

We inherently respond to humanlike forms and gestures, and not only in popular culture. It is interesting to note that neuroscientist have found that specialized parts of the brain are dedicated to recognizing expressions, gestures, and other body-language aspects of human activity. Human body language is understood by many people worldwide, indicating the universality of our body expressions. Therefore, computers may speak to people more effectively if they embody a humanlike form. Movies, shows, advertisements, and many other forms of communication are capitalizing on this human ability to read human gestures and our natural attraction to the humanlike appearance.

Generally the emulation of a human in a machine is a very challenging task. Experience with the development of virtual human actors indicates that the precise duplication of human gestures is extremely difficult. One linguistics study uncovered the depth of the problem for automatic generation of appropriate nonverbal communication. However, the more we work with such robots, the more we understand this technology and the better we can make effective machines that are user friendly.

Besides the mechanical and material aspects of making such robots, there is a need for significant advances in the psychology, neuroscience, and possibly sociological aspects of making these machines. There are many challenges that we may not envision but may be discovered, appreciated, and overcome by making humanlike robots. Constructing a robot that looks and behaves like a human is only one level of the complexity of this challenge. Making robots that are physically functional both in and outside of our homes will involve the need to have them successfully navigate in complex terrains. Some of the obstacles may be stationary, such as stairs and furniture, and others that are dynamic, where the robot will need to move or stop in the path of people, pets, or even automobiles. The complex environment where they will need to walk in a crowded street, crossing a street with traffic while obeying pedestrian laws, walking on sidewalks alongside humans, or even walking on an unpaved road. Such tasks require determining the available path that is safe and within the robot's capability. Achieving these goals is currently the subject of research at many robotic labs worldwide.

Humanlike Robots Help Us Understand and Improve Ourselves

There are many areas where making humanlike robots can help us understand and improve ourselves. On the physical level, the technology is helping to develop better and more realistic prosthetics. On the emotional and psychological level, robots are offering important benefits in the treatment of patients with various phobias, such as fear when speaking in public. Robots that make facial expressions and express themselves verbally using voice synthesizers are already being used to study the potential treatment of children with autism, and reported results are showing great promise. Improving communication skills helps reduce the severity of the disorder of autism, and robots could provide the needed stimulation. Further, as children today are spending more time with computers than with peers of their own age or with people in general, they are growing up with less developed social skills and poorer understanding of body language cues that were taken for granted in prior human generations. This growing concern may be addressed by incorporating humanlike robots into education, therapy, or games while providing realistic simulation that includes controlled conditions. Further details about the use of humanlike robots in health care applications (physical or psychological) and the current state of the art are covered in Chapter 6.

SHOULD WE FEAR HUMANLIKE ROBOTS?

Humanlike robots are being developed to be smart, mobile, and autonomous machines. These capabilities would make them quite powerful. Under certain circumstance, they may also become dangerous to us. Although this technology may improve our lives, it can also cause complications or even terrible destruction if we are not very careful. Some of the concerns may include ethical questions and potential dangers to humans resulting from unlawful acts. To prevent such dangers, we must address the potential concerns long before the possibility of their becoming superior to us is realized.

In order to get the most benefit from their advancing capabilities it is important to channel their development into positive directions and protect ourselves from the negative possibilities. If humanlike robots become more capable and equipped with simulated cognition there will be legitimate concern regarding their continued "loyalty" to us. One may wonder what would happen if they take on questionable roles such as acting as a specific person's clone and then commit a crime, or have access to our assets and private or intimate information and possibly do something to hurt us using our information. Science fiction movies and books are creating public misconceptions of what humanlike robots can do and the danger that they may pose. Yet, as science-fiction ideas are rapidly becoming an engineering reality, it is increasingly becoming important to try to envision the potential issues of concern that may arise and find ways to stave off the possible negative outcomes.

The issues that are associated with the development of humanlike robots involve ethical, philosophical, and religious and safety concerns, and these are all very important to the field's future. Chapter 7 reviews the issues and concerns as well as describes the efforts that are underway to address them.

SUMMARY

Human beings are enormously complex, with systems layered upon systems. It should not be surprising, then, that producing a humanlike robot is highly challenging and involves multidisciplinary tasks that require expertise in many fields, including engineering, computational and material science, robotics, neuroscience, and biomechanics. These technologies are greatly supported by improvements in AI, artificial organs and tissues, artificial vision, speech synthesizers, and so on. As the technology advances, robots will have greater abilities, become more flexible, and possibly be able to fill niche requirements in domestic and industrial applications. Robots may become the first choice in performing critical functions and seeking solutions to future challenges.

Robots are already performing tasks that are dangerous for humans, and they will increasingly perform tasks in areas that are understaffed, such as helping in the care of patients, serving as security guards, guiding blind humans, as well as doing menial or messy jobs that no one wants to do. This may lead us to see them as household appliances or even peers. As they become more pervasive, it may become increasingly important to distinguish them from organic creatures by intentionally designing them with fanciful features that may include a license plate or another visible tagging technique.

Even though the technology today is still too complex and costly and cannot match the performance of special-purpose machines, one can envision that eventually humanlike robots will be designed for military purposes. On the positive side, they may be used as robotic medical staff to treat soldiers on the battlefield. But they may also be used as part of a weapons arsenal, sort of a future army of terminators that may initially be operated remotely and later may be totally autonomous. Such robots will not suffer postwar syndromes and will not have fears and various phobias, as humans do. Although they offer an advantage when used to reduce our military cost in human lives and resources, there is the danger that they will be used in illegal or terror-related operations. Also, robots with artificial cognition that possibly surpass human levels of intelligence may be able to develop a will of their own and potentially turn against people spontaneously.

As the robot's capabilities increase, one may envision the rapid turning of human figures, or models made by computer graphics, into beings that will fill our neighborhoods. With biology inspiring us to move forward with intelligent robotic technology to improve our lives, we will increasingly be faced with challenges to such implementations. Our unconscious fear of this technology may be reduced if we get used to seeing them as a helpful part of our life. Also, it will be beneficial to witness them effectively performing a growing number of unpopular tasks and critical functions. An essential part of accepting having them integrated into our life will be making sure we can avoid potential negative use or accidental acts.

BIBLIOGRAPHY

Books and Articles

Abdoullaev, A., "Artificial Superintelligence," F.I.S. Intelligent Systems (June 1999).

Agrawal, R. N., M.S. Humayun, and J. Weiland, "Interfacing microelectronics and the human visual system," Chapter 17 in Y. Bar-Cohen, (Ed.), *Biomimetics – Biologically Inspired Technologies*, CRC Press, Boca Raton, FL, (Nov. 2005), pp. 427–448.

Arkin, R., *Behavior-Based Robotics*. MIT Press, Cambridge, MA (1989).

Asimov, I., "Runaround" (originally published in 1942), reprinted in *I Robot* (1942), pp. 33–51.

Asimov, I., *I Robot* (a collection of short stories originally published between 1940 and 1950), Grafton Books, London (1968).

Babbage, H., "Babbage's Analytical Engine," a 1910 paper by Henry P. Babbage published in the *Monthly Notices of the Royal Astronomical Society* 70 (1910), 517–526, 645 [Errata].

Bar-Cohen, Y., (Ed.), *Electroactive Polymer (EAP) Actuators as Artificial Muscles – Reality, Potential and Challenges*, 2nd Edition, SPIE Press, Bellingham, Washington, Vol. PM136 (March 2004).

Bar-Cohen, Y., (Ed.), *Biomimetics – Biologically Inspired Technologies*, CRC Press, Boca Raton, FL, (2005).

Bar-Cohen, Y., and C. Breazeal (Eds.), *Biologically-Inspired Intelligent Robots*, SPIE Press, Bellingham, Washington, Vol. PM122 (2003).

Bonde, P., "Artificial support and replacement of human organs," Chapter 18 in Bar-Cohen Y., (Ed.), *Biomimetics – Biologically Inspired Technologies*, CRC Press, Boca Raton, FL, (November 2005), pp. 449–472.

Breazeal, C., *Designing Sociable Robots*. MIT Press, Cambridge, MA (2002).

Capek, K., *Rossum's Universal Robots* (R.U.R.), Nigel Playfair (Author), P. Selver (Translator), Oxford University Press, USA (December 31, 1961).

Cater, John P. *Electronically Speaking: Computer Speech Generation*. Howard W. Sams & Co., Inc., Indianapolis, Indiana (1983).

Dautenhahn, K., and C. L. Nehaniv (Eds.), *Imitation in Animals and Artifacts,* MIT Press, May 2002.

Dietz, P., *People are the same as machines – Delusion and Reality of artificial intelligence*, Bühler & Heckel (2003) (in German)

Drezner, T., and Z. Drezner, "Genetic Algorithms: Mimicking Evolution and Natural Selection in Optimization Models," Chapter 5 in Bar-Cohen Y., (Ed.), *Biomimetics – Biologically Inspired Technologies*, CRC Press, Boca Raton, FL, (November 2005), pp. 157–175.

Fornia, A., G. Pioggia, S. Casalini, G. Della Mura, M. L. Sica, M. Ferro, A. Ahluwalia, R. Igliozzi, F. Muratori, and D. De Rossi, "Human-Robot Interaction in Autism," *Proceedings of the IEEE-RAS International Conference on Robotics and Automation (ICRA 2007)*, Workshop on Roboethics, Rome, Italy, April 10–14, 2007.

Gallistel, C., *The Organization of Action*, MIT Press, Cambridge, MA. (1980).

Gallistel C., *The Organization of Learning*, MIT Press, Cambridge, MA (1990).

Gates, B., "A Robot in Every Home," *Scientific American* (January 2007).

Gould, J., *Ethology*, Norton, (1982)

Hanson, D., "Converging the Capability of EAP Artificial Muscles and the Requirements of Bio-Inspired Robotics," Proceedings of the SPIE EAP Actuators and Devices (EAPAD) Conf., Y. Bar-Cohen (Ed.), Vol. 5385 (SPIE, Bellingham, QA), (2004) pp. 29–40.

Hanson, D., "Robotic Biomimesis of Intelligent Mobility, Manipulation and Expression," Chapter 6 in Bar-Cohen Y., (Ed.), *Biomimetics – Biologically Inspired Technologies*, CRC Press, Boca Raton, FL, (November 2005), pp. 177–200.

Harris, G., "To Be Almost Human or Not To Be, That Is the Question," *Engineering* feature article, (Feb 2007) pp. 37–38.

Hughes, H. C., *Sensory Exotica a World Beyond Human Experience,* MIT Press, Cambridge, MA, 1999.

Kerman, J. B. "Retrofitting Blade Runner: Issues," in Ridley Scott's Blade Runner and Philip K. Dick's *Do Androids Dream of Electric Sheep?* Bowling Green, OH: Bowling Green State University Popular Press (1991).

Lindsay, D. 'Talking head,' *American Heritage of Invention and Technology*, **13**, (1997), pp. 57–63.

Lipson, H. "Evolutionary Robotics and Open-Ended Design Automation," Chapter 4 in Bar-Cohen Y., (Ed.), *Biomimetics – Biologically Inspired Technologies*, CRC Press, Boca Raton, FL, (November 2005), pp. 129–155.

McCartney, S., *ENIAC: The Triumphs and Tragedies of the World's First Computer*, Walker & Company, New York (1999).

Menzel, P. and F. D'Aluisio, *Robo Sapiens: Evolution of a New Species*, The MIT Press, (Sept 2000).

Micera, S., M.C. Carozza, L. Beccai, F. Vecchi, and P. Dario, "Hybrid bionic systems for the replacement of hand functions," *Proceedings of the IEEE*, Vol. 94, No. 9, Sept. 2006.

Mori, M., *The Buddha in the Robot: A Robot Engineer's Thoughts on Science & Religion,* Tuttle Publishing, (1981).

Mori, M., "The Uncanny Valley," *Energy*, 7(4), (1970), pp. 33–35. (Translated from Japanese by K. F. MacDorman and T. Minato)

Musallam, S., B. D. Corneil, B. Greger, H. Scherberger, and R. A. Andersen, "Cognitive Control Signals for Neural Prosthetics," *Science,* 305, (9 July 2004), pp. 258–262.

Mussa-Ivaldi, S., "Real Brains for Real Robots," *Nature,* Vol. 408, (16 November 2000), pp. 305–306.

Perkowitz, S., *Digital People: From Bionic Humans to Androids,* Joseph Henry Press. (2004).

Plantec, P. M., and R. Kurzwell (Foreword), *Virtual Humans,* A Build-It-Yourself Kit, Complete With Software and Step-By-Step Instructions, AMACOM/American Management Association; (2003).

Rosheim, M. *Robot Evolution: The Development of Anthrobotics,* Wiley (1994)

Rosheim, M. "Leonardo's Lost Robot," *Journal of Leonardo Studies & Bibliography of Vinciana, Vol. IX, Accademia Leonardi Vinci* (September 1996): 99–110.

Schodt, F. L., *Inside the Robot Kingdom – Japan, Mechatronics, and the Coming Robotopia,* Kodansha International, New York (1988)

Serruya, M.D., N.G. Hatsopoulos, L. Paninski, M.R. Fellows, and J.P. Donoghue, "Instant neural control of a movement signal," *Nature,* 416 (6877), Mar 14, 2002, pp. 141–142.

Shelde, P., *Androids, Humanoids, and Other Science Fiction Monsters: Science and Soul in Science Fiction Films.* New York: New York University Press (1993).

Shelley, M., *Frankenstein.* Lackington, Hughes, Harding, Mavor & Jones (1818).

Turing, A. M., "Computing machinery and intelligence," *Mind,* 59, (1950), pp. 433–460.

Vincent, J. F. V., "Stealing ideas from nature," *Deployable Structures,* S. Pellegrino (Ed), Springer, Vienna (2005), pp. 51–58.

Walter, W. G. "An Imitation of Life." *Scientific American,* (May 1950), pp. 42–45.

Wessberg, J., C. R. Stambaugh, J. D. Kralik, P. D. Beck, M. Lauback, J.C. Chapin, J. Kim, S. J. Biggs, M. A. Srinivasan and M. A. Nicolelis, "Real-time Prediction of Hard Trajectory by Ensembles of Cortical Neurons in Primates," *Nature,* Vol. 408, (16 November 2000), pp. 361–365.

Wilson, D. H., *How To Survive a Robot Uprising: Tips on Defending Yourself Against the Coming Rebellion,* Bloomsbury Publishing, New York and London, (2005).

Internet Addresses

Beginning of a real robot revolution: giving robots some humanity
    http://www.msnbc.msn.com/id/15831851/
History
    http://bigredhair.com/robots/
    http://chaoskids.com/ROBOTS/directory.html
History of robots
    http://www.businessweek.com/magazine/content/01_12/b3724010.htm
Humanlike robots
    http://www.livescience.com/scienceoffiction/060922_robot_skin.html
    http://www.washingtonpost.com/ac2/wp-dyn/A25394-2005Mar10?language=printer
Karel Čapek
    http://en.wikipedia.org/wiki/Karel_Capek
    http://www.google.com/search?q=robot+pictures&revid=1668943390&sa=X&oi=revisions_in
        line&ct=broad-revision&cd=2
    http://robots.net/
Leonardo's Robot
    http://www.z-kat.com/company/adv_research/leonardo.shtml
    http://www.imss.fi.it/info/eimssdove.html www.imss.fi.it/info/eorgente.html
Robots in Wikipedia
    http://en.wikipedia.org/wiki/Robot
The Human Element: Robots, Robots Everywhere
    http://www.pcworld.com/article/id,117096-page,1/article.html
World's greatest android projects
    http://www.androidworld.com/prod01.htm

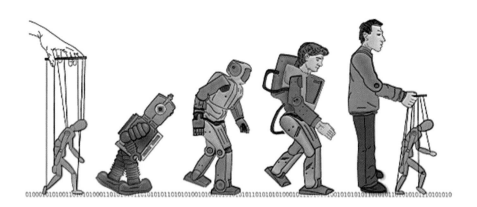

Chapter 2
Emerging Humanoids and Humanlike Robots

In recent years, the number of lifelike humanoids and humanlike robots has grown significantly. Some of the leading countries in which such robots have been developed include Japan and Korea, but researchers and engineers in other countries, including the United States and China, are also extensively involved in developing such robots.

In Japan, besides economic factors and the far-reaching potential of this technology, the need to make humanlike robots is motivated by immediate practical motives, including the major decline in population caused by their record low birthrate and by their having the longest lifespan of any nation (more than 20% of Japan's population is over age 65). As a country with an economy that is one of the strongest in the world, Japan's population decline and aging are raising great concerns regarding the future ability of employers to find employees that will fill low-paying jobs. These jobs may be dangerous, physically demanding, or just unpleasant. Also, the aging population requires a greater number of health care service providers at hospitals and nursing homes. To meet the future need for blue-color employees and service providers, humanoids and humanlike robots are already being developed to look and operate as hospital workers, receptionists, security guards, guides, and so on. Roboticists are crafting these robots to speak in various languages, perform various physical actions (including administering medications and reminding patients to take them), playing musical instruments, dancing to music, and even performing ceremonies.

Generally, these robots range widely in terms of completeness. Some have a head only, some have a head and an upper body without legs, and some are full body.

Y. Bar-Cohen, D. Hanson, *The Coming Robot Revolution*, DOI 10.1007/978-0-387-85349-9_2,
© Springer Science+Business Media, LLC 2009

The mobility of these robots is divided into two basic categories: wheeled locomotion and legged ambulation. Although it is easy for most humans to move on their legs without even thinking about the action, it is quite a challenge to produce two-legged robots that can walk and maintain stability, and it requires an effective control algorithm that is synchronized with the walking mechanism. Another issue that requires attention when making such robots is their safe operation. Robots must be designed such that they will not pose a danger to people or objects along their path; robots must also be protected from potential damage to themselves.

In this chapter, the currently available humanoids and humanlike robots are reviewed to provide a baseline of the state of the art today in terms of available robots and their capability.

HUMANOIDS

Humanoids are robots that are clearly seen as machines but have human characteristics such as a head (with no facial features), a torso, arms, and possibly legs. Generally, these are easier to make than humanlike robots, since their development does not involve the complexity associated with mimicking the appearance and motion of an expressive face. Such robots have already reached high levels of technological maturity, and some of them are even being marketed commercially. Leading Japanese corporations are producing many of these robots, demonstrating their recognition of the potential usefulness of these robots for consumers and industrial users. Several humanoid models are already available for rent and can be seen in shopping centers in Japan. Further, Honda's Asimo robot has been performing in regular stage shows at Disneyland in California in the United States.

HUMANOIDS WITH LEGGED LOCOMOTION

This section describes various examples of humanoids that are ambulated by wheeled locomotion. A summary of the information is given in Table 2.1.

Reborg-Q Security Robots

The security guard company Sohgo Security Services (Alsok) has developed a Robot-Cop that functions as a guard robot. Its commercial name is Reborg-Q^{TM}. This robot moves on wheels and has a head that looks like a helmet. It can be programmed to automatically patrol a preset path or can be remotely controlled by an operator from a central security station. The robot is equipped with four cameras that are mounted in its head and shoulders as well as other sensors that allow for detection of the presence of humans, water, or fire. Whenever one of the robot's sensors is activated, it sends a video to the control room, and the robot issues an alert so that a human operator can take necessary action. The Reborg-Q robot is designed to deter intruders using such tools as a paint gun; it can help put out fire using an extinguisher and can detect water leaks. A touch screen located on the robot's chest can provide information about lost children. Also, a voice synthesizer on the

Table 2.1. Humanoids with wheeled locomotion.

Developer	Humanoid name	DoF and dimensions (height × depth × width)	Weight	Performance and functions
Sohgo Security Services (Alsok)	Robot-Cop	20 DoF with dimensions of 130 × 65 × 70 cm (51 × 26 × 28 inch)	90 kg (200 lb)	Patrols a preset path or remotely controlled by operators from a central security station
Fujitsu	enon	10 DoF with dimensions of 130 × 56 × 54 cm (51.2 × 22 × 21.3 in.)	50 kg (110 lb)	Includes a smart mobile map and speaks in a female voice, greeting visitors and conversing with them.
Mitsubishi	Wakamaru robot	13 DoF with dimensions of 1 m (3.3 ft) tall with a 45 cm (18 in.) circular base	30 kg (66 lb)	Informs people about the weather, performs security patrols, recognizes faces, makes eye contact and then start a conversation, reminds users to take medicine on time, and calls for help if there is a concern that something is going wrong with a patient
NASA	Robonaut	Each of the hands has 14 DoF with 5 in the arms. Its size is an average suited astronaut	Depends on configur-ation	Operated remotely as tele-presence to support EVA such as space walks
NEC	PaPeRo	38.5 cm (15.2 in) tall, 24.8 cm (9.8 in) wide and 24.5 cm (9.6 in) deep	5.0 kg (11 lb)	Serves as a human companion and even as a babysitter; recognizes up to about 650 words. It has batteries that can be operated continuously for approximately 2–3 h

robot can tell the time, provide weather information, and make promotional announcements. A contactless card reader on this robot allows for checking the identification card of employees at company entrances. Alsok is in the process of marketing the Reborg-Q for operation at various commercial facilities in Japan, including shopping malls.

Enon

The enon[TM] (spelled with all lower case letters) was developed as a guide for visitors and also as deliverer of packages (see Figure 2.1). It is made by Fujitsu in Japan and was developed to operate as a smart mobile map that speaks in a female voice. It is capable of greeting visitors at offices and shopping centers, and it has some of the same capabilities as that of Reborg-Q. It is equipped with a touch screen showing maps of various facilities such as ATM machines, public phones, and bathrooms. Its manufacturer, Fujitsu, is seeking to expand its development efforts and begin mass production.

Wakamaru

This robot is named after an ancient samurai warrior, and although it has humanlike eyes, it looks like a machine. It can approach people and inform them about the weather and perform a security patrol of the house when the owner leaves. It can recognize faces, make eye contact, and start a conversation. The robot uses the Linux operating system with multiple microprocessors and can be connected to the Internet. The functions of this robot include reminding patients to take medicine on time and call for help if there is a concern that something bad is happening with a patient who is being monitored. Its manufacturer states that it can detect moving persons and faces, and can recognize two owners and eight other persons. The Wakamaru robot is designed to act spontaneously,

Figure 2.1. The Fujitsu's "enon" robot is being marketed as a service robot to support tasks at offices and commercial establishments. Photo courtesy of Fujitsu Frontech Limited and Fujitsu Laboratories Ltd., Japan.

based on its own preprogrammed activity and daily schedule, which the owner stores in its memory. Its autonomous action depends on three elements: time, place, and behavior. It recognizes its position at its owner's home and moves autonomously, avoiding obstacles according to a movement map around the house. This robot is designed to leave sufficient battery power to be able to return to its charge station for recharging.

The Wakamaru robot was designed to operate in a user-friendly form. One elderly woman who was dying of heart disease requested in her will that her Wakamaru robot attend her funeral! However, Mitsubishi's efforts to market their robot were stopped in 2007. The company is now focusing on making further improvements to the functions of the Wakamaru robot, with the hope of producing a more marketable product.

Robonaut

The Robonaut™ is a tele-operated humanoid that was developed by the Robot Systems Technology branch at NASA's Johnson Space Center in a collaborative effort with the Defense Advanced Research Projects Agency (DARPA). This robot was designed in two key configurations, depending on the application:

1. Operation in space – The robot is mounted on a manipulator arm and has no legs or wheels for mobility (see Figure 1.11).
2. Terrestrial and extraterrestrial operations – A set of two or four wheels is used for mobility (see Figure 2.2).

Figure 2.2. The Robonaut was designed to operate as a robotic astronaut that is tele-operated. It consists of above-the-waist components. Besides the version that is mounted on a robotic arm for space operation (Figure 1.11), it is also designed to move using wheels. Photos courtesy of the Robot Systems Technology at NASA's Johnson Space Center, in Houston, Texas.

The Robonaut is operated remotely by a user wired with sensors, allowing the robot to mirror the user's motion. This robot was developed and demonstrated to function as an equivalent to an astronaut in support of extravehicular activity (EVA), including space walks. The objective of making this robot was to build a machine that could help humans explore space. It is designed to work side by side with humans; the tele-present manipulation of Robonaut allows it to operate in conditions that are too risky for humans while using human intelligence to control the performance.

An important part of the development of Robonaut was to allow it to have dexterous manipulation capability in order to address the reduced dexterity and performance of suited astronauts. Robonaut was designed using ranges of motion, and strength and endurance capabilities, of space-walking human astronauts. The set of EVA tools used by astronauts was the baseline for the initial design of the Robonaut's dexterous five-fingered hand and human-scale arm. The developed robot now has a range of motion that exceeds even unsuited astronauts. The packaging requirements for the entire Robonaut system were derived from the geometry of EVA access corridors, including the pathways on the International Space Station and airlocks built for humans.

PaPeRo

The PaPeRo^{TM} humanoid was developed by NEC to serve as a human assistant, a companion, and even a babysitter. This robot moves on wheels and has two eyes, which gives the robot a cute appearance, but it is far from looking like a real human. It can recognize up to about 650 words of human language and can make its user aware when it has difficulties understanding speech, particularly if it is surrounded by a noisy environment or if the spoken sound is too loud or too soft. When it is not being talked to, it continues to show "signs of life," namely, performing autonomous activities, such as walking around the room, connecting to the Internet to obtain the latest news and updates, or even dancing on its own. These autonomous activities were designed to vary according to the specific robot character. Acting as a babysitter for a group of children, PaPeRo was designed to recognize the faces of individual children and, in case of an emergency, notify their parents of the problem using a cell phone. At the World Expo, which was held in March 2005 just outside the city of Nagoya, Japan, visitors were given the opportunity to leave their children in the care of a PaPeRo robot.

HUMANOIDS WITH LEGGED LOCOMOTION

Bipedal locomotion is much more challenging than mobility on wheels. The related challenges have been addressed in recent years, and significant progress was made in producing robots that can walk using two legs while maintaining stability. Some of the developed robots are described in this section and are summarized in Table 2.2.

Table 2.2. Humanoids with legged locomotion.

Developer	Humanoid name	DoF and dimensions (height × depth × width)	Weight	Performance and functions
Fujitsu	HOAP-2	25 DoF, 50 × 15.7 × 24.5 cm (19.7 × 6.2 × 9.6 in.)	7.0 kg (15.4 lb)	Makes gestures by moving its head, waist, and hands
Honda	ASIMO	34 DoF, 130 × 37 × 45 cm (51 × 14.5 × 17.7 in)	54 kg (119 lb)	Balances itself on one leg, walks forward and backward (including up and down stairs), and turns smoothly without pausing; switches lights on and off and opens and closes doors
JVC	J4 Humanoid	26 DoF, 20 cm (7.8 in.) tall	0.77 kg (1.7 lb)	Centrally controls devices such as home entertainment systems, sings, dances, and acts like it can understand a conversation
Kawada Industries	HRP-3P	36 DoF, 150 cm (4.9 ft) tall	65 kg (143 lb)	Can be operated remotely and autonomously, gets up on its feet from the position of lying down, is designed to work in harsh environments
Korea Institute of Science and Technology (KIST)	Centaur	37 DoF, 1.8 m (5.9 ft) tall	180 kg (397 lb)	Half-humanoid with half-horse mobility platform; designed for human-machine friendly interactions
	Mahru (male) & Ahra (female)	35 DoF, 1.5 m (4.9 ft) tall, about 0.6 m wide (23.6 in)	67 kg (148 lb)	Designed to operate as network-based machine, walks at a speed of 0.9 km/hr (0.56 mph), talks, recognizes gestures, understands speech as well as learns from its own experience
Kitano Symbiotic Systems	Pino	Equipped with 29 motors and is 75 cm (30 in) tall	8 kg (17.6 lb)	Designed as a platform with an open architecture for accelerating the research and development of humanoids
Robo Garage	Female Type (FT) & Chroino (male type)	FT: 23 DoF, 35-cm (13.8 inches) tall	0.8 kg (1.8 lb)	These humanoids are capable of walking like a female or male. The male type bends its body while balancing on one leg
Sony	QRIO	28 DoF (based on the specifications of the SDR-4 model), 0.6 m (2 ft) tall	7.3 kg (16 lb)	Recognizes voice and facial expressions, remembers faces, expresses what it likes and dislikes, stands back up after falling, and

Table 2.2. (continued)

Developer	Humanoid name	DoF and dimensions (height × depth × width)	Weight	Performance and functions
				credited in Guinness World Records (2005 edition) as the first and fastest bipedal robot capable of running
TeamOsaka	VisiON NEXTA	23 DoF, 46.5 cm (18.3 in.) tall, 26 cm (10.2 in.) wide	3.2 kg (7 lb)	Designed to play soccer. Operates autonomously; determines its orientation relative to other objects and selected targets
Tmsuk	Kiyomori	39 DoF, 160 cm (5.25 ft) tall; with its horns reaches 185 cm (6 ft) tall, 70 cm (2.3 ft) wide	74 kg (2.3 lb)	Walks at a speed of 0.5–1.0 km/h (0.3–0.6 mph) and bends and stretches its knees.
Toyota's Partner robots	Walking humanoid Rolling humanoid Mountable humanoid	120-cm (3.9 ft) tall 100 cm (3.3 ft) tall 180 cm (5.9 ft) tall	35 kg (77 lb) 35-kg (77 lb) 75 kg (165.3 lb)	Designed as an assistant for the elderly Moves rapidly for use in manufacturing activities Designed to carry its users into complex terrains
University of Tokyo	Kotaro	91 DoF, 133 cm (4.4 ft) high, 50 cm (1.64 ft) wide and 35 cm (14 in.) thick.	20 kg (44.1 lb) without battery	Walks up a ramp, kicks a ball, and rides a bicycle. It looks in the eyes of people who approach it and shakes their hand
Wow Wee	Robosapien	12 DoF, 61 × 32 × 44 cm (24 × 12.6 × 17.3 in.)	3.6 (8 lb)	Toy that responds to human interaction and environmental stimuli; remotely controlled; walks, turns, bends, sits, waves its hand, stands, lays down, and gets up

HOAP-2

Besides enon, which was described earlier as one of the humanoids that looks like a machine and is mobilized on wheels, Fujitsu has also developed a model that is bipedal. This robot is called HOAP-2^{TM}, and it has two legs and two arms. It makes gestures by moving its head, waist, and hands. The HOAP-2 is a small, simple, and versatile humanoid that was designed as a research tool for scientists and engineers. It can be connected to a personal computer and used as a research tool for studying its movements and operation control as well as the process of communication with humans.

Asimo

Honda's Advanced Step in Innovative Mobility (ASIMO^{TM}) is a two-legged humanoid with very realistic motion. It can walk at a speed of 2.7 km/h (1.7 mph) and run at a speed of 6 km/h (3.7 mph) with an adjustable walking cycle and stride. It has a grasping force of 0.5 kg (1.1 lb) in each of its two five-finger hands and can operate for up to an hour without a battery recharge. The ASIMO robot resembles a small astronaut and has significant agility, including the ability to balance itself on one leg and walk up and down stairs. This humanoid is capable of walking forward and backward, turning smoothly without pausing, and maintaining balance while walking on uneven slopes and surfaces. Also, it can switch lights on and off as well as open and close doors. Honda is seeking to find useful commercial roles for this robot in offices before moving it into the household market. In June 2005, in celebration of the 50th anniversary of Disneyland, an ASIMO robot became a resident robotic entertainer of visitors to the Honda ASIMO Theater in Disneyland in Anaheim, California.

J4 Humanoid

The consumer electronic company JVC is also on the list of major corporations that are developing humanoids. Its researchers and engineers have been developing such robots since the year 2000, seeking to apply the company's strength in miniaturization and mechanisms to design novel robots. The JVC J4 Humanoid^{TM} has an aluminum body frame that can be controlled remotely via cell phone using its Bluetooth capability, but it is also voice activated. It has two video cameras on its head, which operate as the equivalent of eyes and record the videos. Using its internal lithium ion battery the J4 Humanoid can operate for about 90 min without the need for recharge. This robot was also designed to centrally control devices such as home entertainment systems, and it can sing, dance, and act like it can understand a conversation. As many other major manufacturers of humanoids, JVC is not commercializing its robots yet. The current objective is to determine consumers' potential interest while continuing to improve the technology in-house.

HRP-3P

The HRP-3P^{TM} humanoid was developed by Kawada Industries in collaboration with the National Institute of Advanced Industrial Science and Technology (AIST) and

Kawasaki Heavy Industries (KHI). The development of this robot was initially carried out under the Humanoid Robotics Projects (HRP), which was a 5-year program that was launched in 1998 by Japan's Ministry of Economy, Trade, and Industry. The goal was to develop a robot that could work in human environments and use human tools. After the program's completion in 2003, General Robotix, Inc., started commercializing the HRP robots and its building blocks (i.e., servo motors). The HRP-3P robot can be operated remotely and can also work autonomously. This robot uses a flexible torso joint that bends, allowing it to lie down and get back up. It was designed to work in harsh environments and for this purpose it is equipped with effective dust- and splash-resistance.

KIST's Legged Robots

The following legged robots were developed by the Korea Institute of Science and Technology (KIST) in Seoul, Korea.

Mahru and Ahra

Under the lead of Bum-Jae You, KIST developed two models of humanoids; their male model is called Mahru and their female model is called Ahra. These two humanoids were designed to operate as network-based machines that walk at a speed of 0.9 km/h (0.56 mph), talk, recognize gestures, and understand speech as well as learn from their own experience.

Centaur

The Centaur is an autonomous humanoid robot whose name was derived from the Greek half-human/half-horse mythical creature. Its upper body is anthropomorphic, while the mobility is provided by a four-legged platform that carries the robot body. The four legs are driven by hydraulic actuators and the rest of the joints are driven by DC motors. The head is equipped with stereo vision cameras. To allow for friendly human/machine interaction a voice synthesizer is used to speak, which activates the mouth motion for a realistic appearance of conversing. Also, the received sound is analyzed via a voice-recognition system.

Pino

The Pino^{TM} humanoid was developed under the lead of Hiroaki Kitano in Tokyo, Japan, under funding from the government of Japan. The project itself was started at Kitano's company, Kitano Symbiotic Systems, which is a subsidiary of Japan's Science and Technology Corp. Named after Pinocchio, this robot is a two-legged machine with a helmet head and two hands. The robot was designed as a platform with an open architecture, which is intended to accelerate the research and development of humanoids by providing technical information to the general public. This robot participated in the RoboCup events (see later in this chapter) that were held in 2000 and 2001, in which robots played a soccer game.

FT and Chroino

A pair of about 30-cm (11.8-in.) tall robots was developed by the Japanese company RoboGarage. A photo of the FT^{TM} robot, which is one of these two, was shown in Chapter 1 (see Figure 1.3), and both are shown in Figure 2.3. These humanoids were made by the roboticist, Tomotaka Takahashi, and they are capable of walking like a female and male, respectively. The male type, which is called Chroino, is capable of bending its body while balancing on one leg.

QRIO

The Sony humanoid QRIO^{TM} is a child-size robot with a helmet-like head. Its name stands for "Quest for cuRIOsity." It was developed as an entertainment robot that Sony sought to market and was a follow up to its success with the dog-like toy AIBO. Its initial name was Sony Dream Robot (SDR), which represented a series of experimental humanoids. The internal battery of the fourth generation QRIO lasts only about 1 h.

Figure 2.3. The Chroino (*left*) and the Female Type (*right*) presented at the Wired Magazine 2007 NextFest that was held in Los Angeles, California, in September 2007. Photo by Yoseph Bar-Cohen.

QRIO was designed to perform voice and face recognition; it was able to remember people faces as well as express its likes and dislikes. QRIO was capable of running as fast as 23 cm/s (9 in/s), and it was credited in the Guinness World Records (2005 Edition) with being the first and fastest bipedal robot capable of running. The definition that was used for a running robot was "moving while both legs are off the ground at the same time." An additional capability of the QRIO was its standing back up after falling. Generally, recovering from a fall is an inherent problem in many of the humanoids that are bipedal, and the large ones in particular. The larger the robot, the more inertia is involved with its fall and the more damage that occurs. Unfortunately, due to business considerations, on January 26, 2006, Sony canceled the development of this robot. In the experimental marketplace of robotics, not every product is a success, and the QRIO can be considered a loser in this natural selection process.

VisiON NEXTA

In 2003, TeamOsaka was formed in the city of Osaka, Japan, to develop robotic soccer game players seeking to win the Humanoid League prize in the international RoboCup challenge. The team operates under the lead of VSTONE Co., Ltd., and it consists of Systec Akazawa Co., Ltd., Robo Garage, Intelligent Robotic Lab, Osaka University, and the Advanced Telecommunications Research Institute International (ATR). The fully autonomous humanoid that they developed was named VisiON NEXTA^{TM}, and it was the winner of the RoboCup 2004 in Lisbon (for further information about the Robo-Cup challenge see later in this chapter). It has 23 degrees of freedom (DoF) and is equipped with three types of sensors that allow it to determine its orientation relative to other objects. These sensors include (a) an omni-directional camera, allowing the capture of images in 360° (which means it does not need to move its head or turn around to view its surrounding or find targets); (b) a three-axis acceleration sensor that senses the gravitational and external forces in order to recognize whether it is standing or falling; and (c) a three-axis gyro sensor, which provides information about the angular velocity of the robot.

Kiyomori

The Kiyomori^{TM} is a samurai-like robot (see Figure 2.4) that was developed by Tmsuk Co., Ltd., and Atsuo Takanishi Laboratory of Waseda University in Japan. It has the ability to bend and stretch its knees when it walks using the 2 DoF revolving motion in its pelvic region. The Kiyomori humanoid walks at a speed of 0.5–1.0 km/h (0.3–0.6 mph) and uses a NiH battery for power. The Kiyomori has a total of 39 DoF/joints, which enable it to move smoothly. Its first public appearance took place on December 12, 2005, at one of Japan's oldest shrines, Munakata-taisha (the city of Munakata in Fukuoka).

Partner Robots

Toyota's Partner Robots consist of a series of robots that are being developed with human characteristics, which include being agile, friendly, kind, and intelligent enough

Figure 2.4. Kimori is a samurai-like robot that was developed by Tmsuk Co., Ltd., and Atsuo Takanishi Laboratory of Waseda University in Japan. This robot was presented at the Wired Magazine's 2007 NextFest that was held in Los Angeles in September 2007. Photo by Yoseph Bar-Cohen.

to operate devices for personal assistance and care for the elderly. Realizing the need to focus on robots for specialized sets of skills, Toyota is developing three different types of Partner Robots at present, as follows:

- WALKING. The walking humanoid model has two legs and is designed as an assistant for helping the elderly.
- ROLLING. This model is mobilized; its wheels allow it to move quickly, and its compact design helps it to take up minimal space. Since it is part of the related

Toyota's robot series, this humanoid is mentioned here rather than in the section discussing humanoids with wheeled mobility.

- MOUNTABLE. The mountable model is an exoskeleton system that is capable of carrying its user, who controls its movement and direction. This system consists of a chair for carrying a user and is equipped with two legs, allowing the chair to walk. When operating in complex terrains (such as stairs and places with various obstacles), this mountable robot offers significant advantages over wheelchairs.

In order to give their robots the ability to play musical instruments, Toyota developed artificial lips that are claimed to move like a human's lips and hands that are able to play a trumpet like a human. By using an automobile's driving controls Toyota incorporated a stabilizing effect in its robots. Also, using small, lightweight, high-precision sensors, Toyota engineers developed an attitude sensor that detects any tilt of the robots, to help support the robot's stability. Using wires to move the arms and legs, engineers were able to significantly reduce their robots' weight and also increase motion speed.

Kotaro

Professors and students at the University of Tokyo produced the boy-like Kotaro robot with a human-like skeletal structure. Kotaro has two legs, which allow it to walk up a ramp, kick a ball, and ride a bicycle. The Kotaro robot is equipped with sensors that measure the "muscle" length and tension, motor current, temperature, and joint angle (for some of the joints). Also, it has distributed tactile sensors, two eyes, two sound-receiving "ears," and one speaker. One of its features includes looking into the eyes of people who approach it and shaking their hand.

Robosapien

There are many toys that look and move like humans. Some of these were mentioned earlier, including Mattel's doll, 2001 Miracle Moves Baby, and the Hasbro doll called Baby Alive. The Wow Wee's Robosapien is described here as another example of a humanoid robot toy. It is a 61-cm (24-in.) tall humanoid that responds to human interaction and environmental stimuli, as well as to remote-controlled commands. It is designed to perform humanlike body movements and to talk and be programmed. It can pick up objects that are up to about 5 cm (2 in.) in size, including a paper cup, and it can also bend business cards and crumple paper.

TODAY'S HUMANLIKE ROBOTS

The following robots were made by roboticists who were trying to copy the appearance and behavior of humans as realistically as possible. These roboticists are mostly from Japan, Korea, and China, with some from the United States. Examples

Table 2.3. Humanlike robots that are very similar to real humans.

Robot name	Developer	Capabilities and functions
Actroid DER	Kokoro Co., Ltd., Japan	Designed to speak Japanese, English, Korean, and Chinese; replies to 500 questions with a pre-programmed menu of 1,000 answers (including the weather and favorite food); able to choreograph and synchronize motions and gestures
EveR-2 Muse	Korea Institute of Industrial Technology, Korea	Makes facial expressions and performs several dance moves
Geminoid HI-1	ATR Intelligent Robotics and Communication Laboratories, Japan	Sits on a chair and gazes around the room; blinks and fidgets in its seat, moving its foot up and down restlessly. Its shoulders rise gently as though it is breathing; can be operated remotely reproducing the voice, posture, and lip movements of Ishiguro, whom it replicated
Ms. Saya, Cyber-receptionist	Tokyo University, Japan	A female-like robot with an attitude and temper, equipped with voice-recognition of 700 verbal responses, makes facial expressions (i.e., joy, despair, surprise, and rage)
Repliee Q2	Osaka University and Kokoro Co., Ltd., Japan	A female-like robot that exhibits breathing movements, fidgets, makes gestures, blinks, and looks around
Rong Cheng	Chinese Academy of Science in Beijing, China	Greets people in many Chinese dialects; responds to nearly 1,000 words; dances and bows
Zeno	Hanson Robotics, United States	Behaves as a little boy; walks, makes facial expressions, and has eye contact with the person with whom it is conversing. It is being developed to perform facial recognition and operating interactively while learning from its companion/ user
Zou Renti (replica of the originator)	Xi'an Chaoren Sculpture Research Institute, China	Makes very limited movements (turns the head, turns the eyes, and blinks); speaks pre-recorded script or dubbed by a human using a microphone

of developed humanlike robots are described in this section and summarized in Table 2.3.

EveR-2 Muse

The EveR-2 Muse is an entertainer robot (see Figure 2.5), which was designed by the Korea Institute of Industrial Technology. The skin of this robot is made of a silicone material that makes it appear lifelike. It has 60 joints in the face, neck, and lower body that enable the Ever-2 Muse robot to make various facial expressions and perform

Figure 2.5. The EveR-2 Muse is an entertainer robot. Photo courtesy of the Korea Institute of Industrial Technology.

several dance moves. This humanlike robot is 161-cm (5.3-ft) tall, which is the average body size of Korean women in their twenties, and it weighs 60 kg (132 lbs).

Zeno

Zeno is one of the latest humanlike robots (see Figure 2.6). It has many features that allow it to behave as a cute little boy. This robot has a combination of the mobility of the Chroino humanoid (see Figure 2.3) that was made by Robo Garage and a head that has

Figure 2.6. Zeno is currently under development at Hanson Robotics as an interactive learning companion.

boyish appearance, made by Hanson Robotics in the United States. Zeno is able to generate a number of facial expressions and is designed to make eye contact with the person with whom it is conversing. Currently, it is being developed to perform facial recognition and has many other capabilities that are only available in larger humanlike robots. Zeno is being developed as an interactive learning companion, a synthetic pal who has a skin-like facial material called Frubber^{TM} that makes its expressions more realistic while using relatively low power (see Chapter 3 for further details about this skin material). In developing Zeno, one of the developers, who is also a coauthor of this book, was inspired by a book about a troubled robot boy and his quest to find love with flesh-and-blood parents, which was also the source material for the Steven Spielberg's film *Artificial Intelligence: AI*.

Chinese Humanlike Robots

Several models of robots bearing a close resemblance to humans were developed in China. Dion is one of these robots, and it was designed with a female appearance. Further, the Beijing Institute of Technology in China developed several humanlike robots, including the BHR and the Huitong. The BHR robot walks at a speed of 1 km/h (0.61 mph), with a 33 cm (13 in.) distance between steps.

Another Chinese academic institute that has been involved with the development of humanlike robots is the Institute of Automation of the Chinese Academy of Science in Beijing. Its researchers, Yue Hongqiang and his team, introduced in August 2006 a female humanoid that they called Rong Cheng. This robot, which can greet people in many Chinese dialects, responds to nearly 1,000 words, as well as dances and bows, and is on permanent exhibit at the Sichuan Science Museum in Chengdu, where it acts as a receptionist and tour guide.

The Chinese roboticist Zou Renti, from the Xi'an Chaoren Sculpture Research Institute, developed a lifelike robot that he modeled after himself. It was introduced at the 2006 China Robot Expo in Beijing, China. This robot is covered with a silica gel skin that gives it a humanlike appearance. A photo of Zou Renti and his unnamed robot clone are shown in Figure 2.7. This robot uses a DC charge of 6 and 12 V, but it is not

Figure 2.7. The roboticist Zou Renti, China, and his clone robot. Photo by Yoseph Bar-Cohen at the Wired Magazine 2007 NextFest that was held in Los Angeles, in 2007.

really designed to operate under battery power. Its humanlike head is designed to turn around 35°, and its eyes can move from side to side. To make it appear lifelike, it blinks about every 5 s. It is designed to speak a prerecorded script, or it can be dubbed by a human using a microphone.

Japanese Humanlike Robots

Under the lead of Hiroshi Kobayashi, a scientist at Tokyo University of Science, a cyber-receptionist female-like robot (see Figure 2.8) was developed that behaves with an attitude and temper. This robot is equipped with voice-recognition technology allowing the robot to make 700 verbal responses and various facial expressions, including those of joy, despair, surprise, and rage.

In an effort to address deficiencies and limitations in the current cognitive abilities of robots, Hiroshi Ishiguro and his research team are exploring the use of tele-operated robots. Ishiguro called these robots Geminoid and suggested that they need to look like existing persons. An example of such a Geminoid robot is the model HI-1 that looks like Hiroshi Ishiguro himself and was jointly produced by ATR Intelligent Robotics and Communication Laboratories in Japan. Further, he gave the related research its name – android science – which includes the study of bi-directional interaction between robots and humans. This research involves quantitative data concerning the details of robot and human interactions as well as the verification of hypotheses in understanding human nature. In Figure 1.5, a photo was shown of the roboticist Ishiguro and his robot replica Geminoid HI-1. In Figure 2.9, below, another view is given showing Ishiguro and Geminoid HI-1 in an interaction pose, and clearly it is difficult to distinguish between them. The

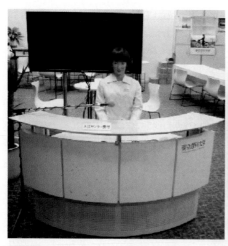

Figure 2.8. The Cyber-receptionist, Ms. Saya, general view (*left*) and close-up (*right*). Photos courtesy of Hiroshi Kobayashi, Tokyo University of Science.

Figure 2.9. The roboticist Hiroshi Ishiguro (*right*) next to the robot Geminoid HI-1 that has his own image (*left*). Photo courtesy of Hiroshi Ishiguro, ATR Intelligent Robotics and Communication Laboratories.

Geminoid HI-1 acts in a humanlike fashion. It sits on a chair and gazes around the room, blinks and fidgets in its seat, moves its foot up and down restlessly, and gently raises its shoulders as though it were breathing. It can also be operated remotely reproducing the voice, posture, and lip movements of Ishiguro.

Ishiguro has been involved with the development of another well-known humanlike robot. This female humanlike robot is named Repliee and was developed jointly by researchers and engineers from Osaka University (Ishiguro Lab) and Kokoro Co., Ltd. The model Repliee Q2 was shown in Chapter 1 (Figure 1.4). Also, below in Figure 2.10, Ishiguro is shown interacting with this humanlike robot. The previous design, Repliee Q1-expo, was modeled after the Japanese newscaster Ayako Fujiia from the TV station NHK. To make the face of the Repliee Q2 model, faces of several young Japanese women were scanned, and an average face image was formed to suggest the appearance of an anonymous young woman of Japanese descent.

The Repliee Q2 robot is made with a soft and colored 5-mm (∼0.2 in.) thick silicone skin that covers its metal skeleton. It is constructed with moist lips, glossy hair, and vivid eyes that blink slowly. It can sit on a chair with hands folded neatly on its lap, and it wears a bright pink blazer and gray pants. There are 42 actuators in the robot's upper body, which are powered by an air compressor, and the Repliee Q2 is programmed to make her movements look human. Her head can move in nine directions, and she makes gestures with her arm, but she is mostly designed to sit. Four high-sensitivity

Figure 2.10. The "female" named Repliee Q2. Photo courtesy of Hiroshi Ishiguro, Osaka University (Ishiguro Lab) and Kokoro Co., Ltd.

tactile sensors are mounted under the skin of this robot's left arm to allow it to react in various ways to different pressures.

The Actroid DER is a female guide robot that was initially developed by Kokoro and later by Osaka University and Advanced Media, Inc., which specializes in speech technology. The company Kokoro is part of the Sanrio Group, which specializes in the design and manufacture of robots. The Actroid DER was designed to speak Japanese, English, Korean, and Chinese and to reply to 500 questions with a pre-programmed menu of 1,000 answers. Some of the topics of conversation that this robot is programmed to handle include the weather and favorite foods. When Actroid DER is asked a question it cannot handle it tries to change the subject of the conversation or encourages others to speak. This robot has a wide repertoire of expressions, but it has problems interpreting speech even in mild crowd noise. Its limbs, torso, and facial expressions are controlled by a system of actuators powered by pneumatic pressure. Once programmed, the Actroid is able to choreograph and synchronize motions and gestures with its own voice. Currently, the manufacturer Kokoro is offering to rent their Actroid DER2 to be used for presentation at various events.

HUMANLIKE ROBOT COMBOS

Most humanlike robots that have been described so far are either clearly identifiable as machines or are as close a copy to humans as possible. However, at least one group of researchers has tried to create a combination of the two. Its robot, called Hubo (short for "humanoid robot"), is a bipedal humanoid. Hubo was developed by the Korea Advanced Institute of Science and Technology. Hubo has voice recognition and an imaging system that uses two eyes. It is quite limited in its mobility; it cannot walk up or down stairs, nor can it run. However, its separated fingers allow it to perform functions that some of the other sophisticated robots, such as the ASIMO, cannot do, including play finger games such as rock, paper, and scissors. The developed prototype, which combined the Hubo body and a head modeled after Albert Einstein, was named "Albert Hubo." It can operate continuously for about an hour without the need for recharging its batteries.

ROBOTS WITH HUMANLIKE HEAD

The most complex part of humanlike robots is the head. Although most roboticists are focusing on making the full body of a robot, some roboticists have focused their efforts on making heads that are fully expressive. Some details about the Kismet robot head that was made by Breazeal were given in Chapter 1. At the 2005 NextFest Conference that was held in Chicago, Illinois, an expressive head of the science-fiction writer Philip K Dick (see Figure 2.11) was presented. This popular U.S. writer has had many of his science fiction novels turned into popular movies, including *Blade Runner, Minority Report,* and *Total Recall.* The humanlike robot that was made to emulate this writer speaks and makes facial expressions as well as gestures.

OTHER HUMANOIDS AROUND THE WORLD

Besides Japan, Korea, China, and the United States, humanoids and humanlike robots are being developed in many other countries. Currently, most of these robots are made as humanoids, and the majority of them are at the stage of models and prototypes. The countries in which such humanoids are being made are as follows.

Figure 2.11. Expressive robotic head of the science fiction writer Philip K Dick. This head was produced by Hanson Robotics.

Australia

The Australian robot Johnny Walker is a 40-cm tall humanoid developed by scientists at the University of Western Australia. This robot has 4 DoF in each of its legs and force sensors in the feet. Also, it is equipped with vision cameras, two accelerometers, and two gyroscopes to provide position, stability, and feedback. This robot was used to play soccer at the 2000 RoboCup competition (see later in this chapter). The Mobile Robotics Laboratory of the School of Information Technology and Electrical Engineering at the University of Queensland in Brisbane, Australia, developed the GuRoo humanoid. It is 1.2 m (4 ft) tall, weighs 38 km (84 lbs), and has 23 DoF. It is a bipedal humanoid that can operate autonomously. The GuRoo was developed for studies of dynamic stability, human–robot interaction (HRI), and machine learning. The GuRoo participated in the annual RoboCup that was held in Japan in 2002. A privately produced humanoid made in Australia is the Tron-X robot that was made by Fesco. This robot was developed in 1997, is 2.8 m (9.2 lb) tall, and weighs about 300 kg (660 lbs). It is actuated by over 200 pneumatic cylinders, which allow it to perform many functions and tasks.

Austria

The BarBot is a humanoid that was developed by the Humanoid Robotics Laboratory (HRL) of Linz, Austria. This robot has a head, torso, and hands, but its hand does not move and the robot is not mobile except for being supported by a set of wheels that can be pushed to different locations. The BarBot is 1.7 m (5.6 ft) tall, weighs 30 kg (66 lb), has 7 DoF, and some other quite limited capabilities. It is designed as an anthropomorphic machine; namely, it has some of the attributes of humans in an animated machine. This robot was designed to ask for money at a bar, hold a beer can, and pour beer into its system container, emulating drinking.

Bulgaria

A humanoid is being developed by a company called Kibertron, Inc., which is located in the city of Sofia, Bulgaria. The robot is called by the same name as the manufacturer, Kibertron. This humanoid looks like a terminator; it is 1.75 m tall (5.7 ft) and weighs 90 kg (198 lbs). Each of its full arms has a total of 28 DoF and each of its hands has 20 DoF.

England

Development of humanoids in England is carried on in both academia and industry. Since 1987, the Shadow Robot Company in London, England, has been developing a humanoid bipedal walking machine. The focus of this company's efforts is on the torso and legs (see Figure 2.12), where 28 pneumatic actuators are used to activate eight joints, enabling a total of 12 DoF. These actuators are used to mimic human walking, but there are significantly fewer actuators on the robot than there are muscles in the

Figure 2.12. Humanoid bipedal walking machine. Photo courtesy of the Shadow Robot Company in London, England.

human leg. The mounting of the actuators at certain attachment points, making them capable of withstanding the exerted strains (not identical to the strains placed on human muscles), determined the placement of the anchor points on the robot skeleton.

At the Bristol Robotics Laboratory (BRL), University of Bristol, and the University of the West of England in cooperation with Elumotion Ltd. in Bath, the humanoid BET-1 (see Figure 2.13) is being developed. This robot is intended to address some fundamental issues related to safe HRI, specifically, the issues that arise when a human

Figure 2.13. The BET-1 and the leader of its development team, Chris Melhuish. Photo courtesy of Chris Melhuish, Bristol Robotics Laboratory in England.

and a robot must perform cooperative tasks in a confined space. Such conditions may include performing joint tasks in the kitchen, helping an artisan on a building site; or even assisting a surgeon during an operation. Issues related to HRI safety may involve communication of a shared goal (verbally and through gesture), perception and under-standing of intention (from dexterous and gross movements), cognition necessary for interaction, and active and passive compliance, as well as the social behavior of 'affective robotics.' These are important issues for many applications of humanoids, and these researchers are seeking to establish the scientific foundations from which to engineer cognitive systems on a wide scale. The BET-1 has 17 DoF, including 2 DoF for the neck joint, 3 for the shoulder, 1 for the elbow; 2 for the wrist; and 9 for the hand. Currently, the robot is faceless, but there are plans to incorporate facial expressions as well as body gesture recognition and generation. However, the team has no current plans to add legs for mobilizing the robot.

Further, researchers at the Department of Electrical and Electronic Engineering, the Imperial College of London, are working on an upper torso (bench-mounted) of a humanoid called Ludwig. This robot has two arms, each with 3 DoF, and a stereoscopic camera mounted on a pan-and-tilt head. Ludwig is used for fundamental research in cognitive robotics, with the objective of equipping robots with the capability of high-level cognition. To simplify the control problems and concentrate on the issues of perception, knowledge representation, and cognition this robot was designed with fewer degrees of freedom than many other humanoids.

European Consortium

The European Commission has funded the RobotCub Consortium, which consists of several European universities under the coordination of the University of Genoa in Italy. One goal of this consortium is to develop the humanoid called iCub to replicate the physical and cognitive abilities of a two and a half year old child. This robot is being developed as an open platform for studying cognition through the implementation of biological motivated algorithms. The developed software is produced as an open source for use by other partners to establish worldwide collaboration. Efforts are being made to develop the iCub with an attention system that uses a combination of visual and auditory cues to allow interactive communication as well as the ability to reach out and grasp visually identified objects.

In the initial design, the iCub was 90 cm tall and weighed <23 kg. It was equipped with 53 DoF (see Figure 2.14). Recently, the iCub was redesigned into a 104-cm tall robot with a simplified leg structure. Prelinguistic cognitive capabilities that were already built into this robot include seeing, appendage movements, and imitation. The RobotCub Consortium is also involved with a project called ITALK (Integration and Transfer of Action and Language Knowledge in Robots). This project is under the lead of the University of Plymouth in England, and its objective is to use the iCub to study how a robot might quickly pick up language skills.

Germany

In Germany, there are several academic and research institutes that are developing humanoids. The robot Care-O-Bot was developed by scientists at the Fraunhofer Institute, and it was designed as a mobile service robot that is capable of interaction with and assisting humans in typical housekeeping tasks. This robot looks somewhat like R2D2 in the *Star Wars* movies, and it was designed to help maintain independent lifestyles by people who are elderly, disabled, or in rehabilitation. The Care-O-Bot was developed to guide and support people who need assistance when they are not steady on their feet. Some of its functions include operating home electronics and performing various tasks around the house. Three such robots were installed in March 2000 for permanent display at the Museum für Kommunikation Berlin, where they autonomously move among the visitors and communicate with them while avoiding collision.

At the Forschungszentrum Informatik (FZI), the University of Karlsruhe in Germany, a humanoid called ARMAR is being developed. This robot consists of an autonomous mobile wheel-driven platform, a body with 4 DoF, a two-arm system with a simple gripper, and a stereo camera head. The total weight of ARMAR is about 45 kg

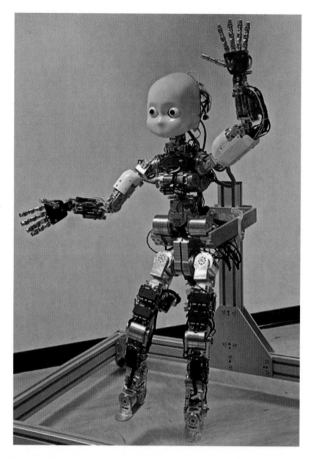

Figure 2.14. The development of the iCub humanoid robot was funded by the European RobotCub Consortium. Photo courtesy of the RobotCub Consortium.

(100 lbs), and the robot consists of two actively driven wheels fixed in the middle of an octagonal board and another two wheels as passive stabilizers. The maximum velocity of its platform is about 1 m/s, and it can rotate about 330°. The ARMAR can bend forward, backward, and sideways. To increase the height of this robot, a telescopic joint is included in its body, allowing it to get up to 40 cm (15.7 in.). ARMAR has two arms; each has 7 DoF and is 65 cm (2.1 ft) long. To work in cooperation with humans, the physical structure (dimension, shape, and kinematics) of each of its arm was developed to resemble the human arm in terms of segment lengths, axis of rotation, and workspace. In order to make the robot a helpful assistant to humans in everyday life, the researchers at FZI are seeking to make it able to learn and cooperate with humans.

The Department of Computer Science at the Technical University of Darmstadt in Germany developed a humanoid called Lara (see Figure 2.15). According to its developers' website, the name refers to the character in the book and movie *Doctor Zhivago*

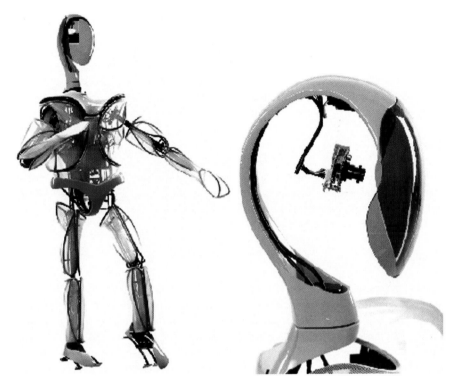

Figure 2.15. A full view and the head of the humanoid Lara. This robot is actuated by shape memory wire bundles, allowing significant weight reduction compared to robots with electric motors. Photo courtesy of Robert Kratz, Technical University of Darmstadt/HFG of Main, Germany.

(1965). Lara is 130 cm (4.3 ft) tall, weighs 6.5 kg (14 lbs) and is equipped with 34 Nitinol shape memory alloy wire-bundled actuators in support of its 18 DoF. The development of Lara was done under the lead of Oskar von Stryk. The use of shape memory alloys allows this robot to have a reduced weight equaling about 1/6 that of comparable robots that are driven by rotary motors.

At the Institute of Applied Mechanics, Technical University of Munich (TUM), a humanoid called Lola is being developed that is equipped with 22 DoF, where each of the legs has 7 DoF, the upper body has two, and each of the arms has four. This robot is being developed as part of a project called "Biological and Technical Aspects of Intelligent Locomotion," which is funded by the German Research Foundation. The developers of this robot are hoping to enable their robots to walk fast.

Italy

Researchers under the lead of Guiseppe Menga at the Politecnico di Torino have developed a humanoid called Isaac. This robot is 85 cm (2.8 ft) tall, weighs about

15 kg (33 lb), and has 16 DoF. Isaac is equipped with 6 DoF in each of its legs, allowing the legs to perform three-dimensional movements. It has two motors in the neck, allowing rotation of the cameras and making the robot able to track a ball in real time. In its prototype version the arms have only 1 degree of freedom. Its computer is equipped with three software modules to allow artificial vision and sensor processing, as well as game planning and walking. This humanoid won the second place prize in the Robocup 2003.

Russia

In 2003 a St. Petersburg's company called New Era and students from the St. Petersburg Polytechnic University completed a 2-year project in which two humanoids called ARNE and ARNEA were developed. The height of these robots is 1.23 m (4 ft), they weigh 61 kg (134 lbs), and they are equipped with 28 DoF.

Singapore

Scientist and engineers at the Advanced Robotics and Intelligent Control Centre (ARICC) of the School of Electrical and Electronic Engineering at the Singapore Polytechnic are also developing humanoids. Their robots are identified as Robo-Erectus (RE), and the latest one is RE80II. This humanoid is 80 cm (2.6 ft) tall, weighs 7 kg (15 lbs), and has 22 DoF. It is equipped with accelerometers, force sensors, gyros, and range sensors as well as stereo cameras. Its development research team is seeking to make its low-cost soccer-playing humanoids capable of walking, turning, crouching, and kicking a soccer ball past a goalkeeper into the goal.

Spain

Under the lead of Davide Fraconti, Pal Technology, in Barcelona, Spain, a humanoid is being developed called REEM_A. This robot is 1.4 m (4.6 ft) tall, weighs 41 kg (90 lbs), has 30 DoF, and can operate autonomously for about an hour and a half without recharging its batteries. REEM_A has a walking speed of 1.5 km/h (0.93 mph) and is equipped with the ability to perform simultaneous localization and mapping through visualization. It can recognize various objects that are presented to it (see Figure 2.16). It can play chess autonomously and has voice and face recognition capability. It can grasp objects that weigh up to 6 kg (3.7 lb). Also, it can perform tele-presence tasks via an Internet connection. This robot was introduced at the Robocup competition in 2006, where it won the foot race competition.

United States

There are several research and development organizations in the United States where humanoids are being developed, mostly with heads that have no facial features and only the general shape of a human. Some of these robots were developed to operate as a human assistant; an example includes the Nursebot that is described in the next section.

Figure 2.16. The humanoid REEM_A (made by Pal Robotics in Spain) can recognize objects that are presented to it. Photo by Yoseph Bar-Cohen at the Wired Magazine 2007 NextFest that was held in Los Angeles, California, in September 2007.

Many of the organizations where such robots are developed in the United States are academic institutes.

For example, the Laboratory for Perceptual Robotics (LPR) of the Computer Science Department, University of Massachusetts, developed an upper half of a humanoid called Amherst. This humanoid has 29 DoF distributed in two arms each with 7 DoF and two three-fingered hands with 4 DoF each. The lab studies computational systems that control sensory and motor abilities, and the humanoid Amherst was developed to support this study. The robot is effectively an experimental platform that includes sensor networks and actuators for mobility and manipulation. The focus of this lab is on dexterity of robots, force-guided manipulation, knowledge representation, HRI and communication as well as computational models for learning and development.

At the Massachusetts Institute of Technology (MIT) a set of legs called M2 and an anthropomorphic torso called COG were developed as important parts of humanoids. COG includes a head with four eyes (two for close up and two for distance) and two arms. The M2 is a 3D bipedal walking robot that can walk at a speed of 1 m/s and has 12 active DoF – 3 in each hip, 1 in each knee, and 2 in each ankle.

The Center for Intelligent Systems at the School of Engineering, Vanderbilt University, is developing a humanoid called ISAC. This humanoid was designed and built as a research platform for the development of assistive robots in support of elderly in-home care. It is equipped with robot-to-human communication tools, including audio, visual, and gestural components. The ISAC robot is also equipped with two sets of 6 DoF soft arms that are actuated by McKibben artificial muscles driven by an air compressor system. The hands contain four fingers that operate as anthropomorphic dexterous manipulators and are supported by two force–torque sensors that are connected at the arms' wrist joints.

Industry is also starting to play a role in developing such robots. An example of the humanlike toy Zeno that was developed by Hanson Robotics was described earlier. Another example is Faustex Systems Corporation. The robot developed by Faustex Systems is called Hyperkinetic Humanoid (H2-X), and it is an artistic robot with anthropomorphic features. According to its developers, this machine is faster than humans, and it was designed to operate safely when it is physically interacting with humans. This robot is 193 cm (6.3 ft) tall, weighs about 163 kg (300 lb), and has 24 DoF. The H2 humanoid is covered with lightweight foam and fabric.

ROBOTS IN HEALTH CARE

One of the most important areas for which humanoids and humanlike robots are being developed is health care, particularly for assisting people who need help or medical treatment. Robots are developed to provide therapy to the elderly who, in some countries such as Japan, are filling nursing homes at an alarming rate and many times suffer depression and loneliness. Besides helping to reduce stress and depression among the elderly, robots are being designed to provide patients and caregivers with as much support and assistance as possible. Robots with humanlike characteristics have been developed to operate as nurses; one example is the 122-cm (4-ft) tall, 34-kg (75-lb) Nursebot that was developed by Carnegie Mellon University (CMU) jointly with the University of Pittsburgh. This robot was developed under a grant from the National Science Foundation (NSF), and two prototypes were produced named Pearl and Florence. The Nursebot was mounted on a platform with wheels that allows for its mobility. This humanoid was designed to remind patients to take their medication on time and to alert the medical staff about the needs of patients in an emergency situation. The acquired information is communicated electronically using the Internet and video

Figure 2.17. Carnegie Mellon's Nursebot Pearl at Longwood Gardens in Pittsburgh. Photo courtesy of Sebastian Thrun, Stanford University.

conferencing. A photo of the Nursebot Pearl in the hospital setting of Longwood Gardens, Pittsburgh, is shown in Figure 2.17.

Another example of a robot that was designed to assist caregivers is the RI-MAN (Figure 2.18), which was developed by the Biomimetic Control Research Center of RIKEN in Japan. The RI-MAN robot is able to carry a manikin, substituting as a real patient, picking it up from one bed and moving it to another one, and doing this task in a similar way to a human nurse. To be safe when operating in physical contact with humans the whole body of the RI-MAN robot is covered with soft material, and the mechanical joints are physically isolated. The RI-MAN's forearms, upper arms, and

Figure 2.18. The RI-MAN robot carries a manikin, who simulates a patient. Courtesy of RIKEN Bio-Mimetic Control Research Center, Japan.

torso are equipped with soft tactile sensors that measure the magnitude and position of the contact forces. Feedback of the tactile information from these sensors is used by the RI-MAN operating system to allow for physical interactions with humans without causing harm. Each of the RI-MAN's two arms has six joints that are driven by six motors and are operated in pairs to enable combined bending and twisting motions. The RI-MAN's arms use a mechanism that provides the shoulder and elbow power when lifting and holding a human.

DEVELOPMENT THROUGH COMPETITION

Competition has been an effective way to promote and encourage rapid advances in the technologies needed for humanlike robots. There are several annual robot contests. One of these is the RoboGames (previously known as ROBOlympics) that are held in San Francisco, California, and were founded by David Calkins in 2004. The objective was to provide a forum for cross-pollination between events and to address the fact that too many robot builders are over-specializing within their own field. The idea behind bringing together designers from combat robotics (mechanical engineering), soccer robotics (computer programming), sumo robotics (sensors), android robotics (motion control), and art robotics (aesthetics) is to have robot builders exchange ideas and learn from each other.

Another competition, which was mentioned earlier in this chapter, is the RoboCup. This competition represents an international challenge to promote advances in robotics, artificial intelligence, and related fields. It involves teams of multiple fast-moving robots that are operating under a dynamic environment. The ultimate goal of the RoboCup challenge is to develop by 2050 a team of fully autonomous humanoids that can win in a soccer game against a human world champion team. To reach this objective there is a need to incorporate various technologies that include high speed computing, autonomy, collaboration, strategy acquisition, real-time reasoning, sensor-fusion, and many other disciplines.

SUMMARY

There is an impressive list of robots with human characteristics ranging from humanoids that look like humans but with helmet-like heads to humanlike robots that have a very similar appearance to humans. Some of the developed robots are produced by major corporations such as Honda, NEC, and Toyota, and these are the humanoid types. The robots that are mobilized on platforms with wheels are the closest in maturity to becoming commercial products. Some of the humanoids are already available for rent, allowing consumers to interact with the robots and provide feedback to the robot producers. Such robots are already showing up in Japan in various shopping centers and malls. Humanlike robots have been made primarily in Japan, Korea, China, and the United States, and they exhibit movements, facial expressions and gestures that are very human. Several roboticists have even made robots that are clones of themselves.

The existing humanoids and humanlike robots are a long way from where we want to be. They are designed to perform limited functions. These limitations make them less attractive for wide use in the near future. They are also still too expensive and have batteries that need to be recharged after only an hour or an hour and a half. Of notable concern is the fact that after years of development of its highly visible experimental robots in the SDR series, Sony canceled the development of its ORIO robot, which was supposed to become one of its commercial successes.

BIBLIOGRAPHY

Books and Articles

Aldiss B. W., *Supertoys Last All Summer Long: And Other Stories of Future Time, St. Martin's Griffin,* (June 27, 2001).

Beira R., M. Lopes, M. Praca, J. Santos-Victor, A. Bernardino, G. Metta, F. Becchi and R. Saltaren, "Design of the Robot-Cub (iCub) head," *Proceedings of the IEEE/RSJ Int. Conf. on Intelligent Robots and Systems,* Orlando, FL, (2006) pp. 94–100.

Chou C.P., B. Hannaford, "Measurement and modeling of Mckibben pneumatic artificial muscles," *IEEE Transactions on Robotics and Automation,* Vol. 12, (Feb. 1996), pp. 90–102.

Chou, C. P., and B. Hannaford, "Static and dynamic characteristics of Mckibben pneumatic artificial muscles," *Proceedings of the IEEE International Conference on Robotics and Automation,* Vol. 1, San Diego, CA, (May 8–13, 1994), 281–286.

Dunn E., and R. Howe, "Foot placement and velocity control in smooth bipedal walking," *Proceedings of the IEEE Int. Conf. Robotics and Automation,* (1996) pp. 578–583.

Hans, M., "The Control Architecture of Care-O-bot II." In: E. Prassler, G. Lawitzky, A. Stopp, G. Grunwald, M. Hägele, R. Dillmann, and I. Iossifidis, (Eds.): *Advances in Human-Robot Interaction Series: Springer Tracts in Advanced Robotics,* Vol. 14 (2004), pp. 321–330.

Hirai K., M. Hirose, Y. Haikawa, and T. Takenaka, "The development of Honda humanoid robot," *Proc. IEEE Int. Conf. Robotics and Automation,* (1998).

Hirose M., Y. Haikawa, T. Takenaka, and K. Hirai, "Development of humanoid robot ASIMO," *Proceedings of the International Conference on Intelligent Robots and Systems,* Maui, HI, workshop 2 (2001).

Kajita, S., and K. Tani, "Experimental study of biped dynamic walking in the linear inverted pendulum mode," *Proceedings of the IEEE Int. Conf. Robotics and Automation,* (1995).

Kaneko K., F. Kanehiro, S. Kajita, H. Hirukawa, T. Kavasaki, M. Hirata, K. Akachi and T. Isozumi, "Humanoid robot HRP-2," *Proceedings of the IEEE Int. Conf. on Robotics and Automation,* New Orleans, LA, (2004), pp. 1083–1090.

Kazerooni, H., and J. Guo, "Human Extenders", *Transactions of the ASME, Journal of Dynamic Systems, Measurements, and Control,* Vol. 115, (1993) pp. 281–290.

Kratz R. M. Stelzer, M. Friedmann and O. von Stryk, "Control approach for a novel high power-to-weight ratio SMA muscle scalable in force and length," *Proceedings of the IEEE/ASME Intl. Conf. on Advanced Intelligent Mechatronics (AIM),* Zürich, Switzerland, (September 4–7, 2007)

Lohmeier, S., T. Buschmann, H. Ulbrich, and F. Pfeiffer, "Modular Joint Design for a Performance Enhanced Humanoid Robot," *Proceedings of the IEEE International Conference on Robotics and Automation (ICRA),* (2006).

McGeerr, T., "Passive dynamic walking," *International Journal of Robotics Research,* Vol. 9, (1990) pp. 62–82.

Metta, G., G. Sandini, D. Vernon, D. Caldwell, N. Tsagarakis, R. Beira, J. Santos-Victor, A. Ijspeert, L. Righetti, G. Cappiello, G. Stellin and F. Becch, "The RobotCub project an open framework for research in embodied cognition," *Invited paper in the Proceedings of the International Conference of Humanoid Robots,* Workshop on Dynamic Intelligence, Tsukuba (2005).

Miura, H., and I. Shimoyama, "Dynamic walk of a biped," *International Journal of Robotics Research*, Vol. 3, No. 2, (1984) pp. 60–74.

Mizuuchi I., T. Yoshikai, Y. Nakanishi, Y. Sodeyama, T.Yamamoto, A. Miyadera, T. Niemelä, M. Hayashi, J. Urata, and M. Inaba, "Development of Muscle-Driven Flexible-Spine Humanoids," *Proceedings of Humanoids 2005, part of the 5th IEEE-RAS International Conference on Humanoid Robots* (December, 2005) pp. 339–344.

Nishio, S., H. Ishiguro, and N. Hagita, "Geminoid - Teleoperated Android of an Existent Person," *Humanoid Robots: New Developments*, A.C. de Pina Filho (Ed.), Tech Education and Publishing, Vienna, Austria, (2007).

Pratt, G. A., "Legged Robots at MIT: What's new since Raibert," *Research Perspectives, IEEE Robotics and Automation Magazine*, (September, 2000) pp. 15–19.

Raibert, M., *Legged Robots that Balance*, Cambridge, MA: MIT Press (1986).

Yamaguchi, J., E. Soga, S. Inoue, and A. Takanishi, "Development of a bipedal humanoid robot – Control method of whole body cooperative dynamic biped, walking," *Proceedings of the IEEE Int. Conf. Robotics and Automation*, (1999) pp. 368–374.

Internet Websites

Actoid DER http://www.kokoro-dreams.co.jp/english/index.html

Actroid http://en.wikipedia.org/wiki/Actroid

Android world http://www.androidworld.com/

ASIMO is now residing at Disneyland
http://asimo.honda.com/NewsArticle.aspx?XML=News/newsarticle_0034.xml

BarBot robot (made by HRL, Austria) http://www.roboticslab.org/html/index.php

Carnegie Mellon contributions to walking Robots
http://www.newdesign.cs.cmu.edu/archive/2007/pages/RHOFReleasePhotos.pdf

Chalmer's Humanoid Project http://humanoid.fy.chalmers.se/links.html

Chinese Humanoid Robot http://www.techshout.com/science/2006/26/chinese-blond-female-humanoid-robot-unveiled/

Desktop Robot Mascot - JVC J4 http://www.robots-dreams.com/2006/01/desktop_robot_m.html

EveR-2 Muse -
http://www.aving.net/usa/news/default.asp?mode=read&c_num=31719&C_Code=07&SP_Num=0

Federation of International Robotic-Soccer Association http://www.fira.net/

Female Guide Robot http://www.plasticbamboo.com/2006/10/05/female-guide-robot-from-sanrio/

Game machines play (including World cup) http://www.pbs.org/saf/1208/resources/resources-1.htm

Gynoid (Female-like robots) http://en.wikipedia.org/wiki/Gynoid

Hanson Robotics http://hansonrobotics.com
http://hansonrobotics.com/movies/sci_ch_NeXtFesT.asf

Honda's Asimo http://asimo.honda.com/

Humanlike robots http://www.flashedblogs.com/filedunder/robots/193.html?page=6

Humanlike robots http://www.livescience.com/scienceoffiction/060922_robot_skin.html
http://www.washingtonpost.com/ac2/wp-dyn/A25394-2005Mar10?language=printer

Humanoid & Anthromorphic Robots http://www.service-robots.org/applications/humanoids.htm

Humanoid robots http://en.wikipedia.org/wiki/Humanoid_robot

Humanoid robots http://www.droidlogic.com/

Humanoid robots in Wikipedia http://en.wikipedia.org/wiki/Humanoid_robot

Humanoids http://www.businessweek.com/magazine/content/01_12/b3724007.htm

Japanese develop 'female' android http://news.bbc.co.uk/1/hi/sci/tech/4714135.stm
http://www.kurzweilai.net/news/frame.html?main=news_single.html?id=5995

Japan's Love Affair with Androids
http://www.msnbc.msn.com/id/14270827/site/newsweek/page/3/

JSK Lab, Dept. of Mechano-Information, Graduate School of Information Science and Technology, University of Tokyo http://www.jsk.t.u-tokyo.ac.jp/research.html

Kilberton http://www.kibertron.com/

KIST humanlike robots http://humanoid.kist.re.kr/eng/mahruahra/show_01.php

Kotaro http://www.jsk.t.u-tokyo.ac.jp/~ikuo/kotaro/index-e.html

NEC's PaPeRo Robot http://www.incx.nec.co.jp/robot/english/intro/intro_01.html

Pino robot http://www.symbio.jst.go.jp/~tmatsui/pinodesign.htm

Robo Sapiens http://www.designboom.com/eng/education/robot1.html

RoboCup http://www.robocup.org/

RoboGames (formerly Robolympics) http://en.wikipedia.org/wiki/Robolympics

Robosapien http://216.176.50.35/robosapien/robo1/robomain.html

Robotcub and iCub http://www.robotcub.org/misc/review3/07_Tsagarakis_et_al_JAR.pdf

Robot books http://www.robotbooks.com/

Robots and Androids http://www.vaasapages.com/RobotsAndroids.htm

Robotics and Automation Society, Service Robots http://www.service-robots.org/applications/humanoids.
    htm

Robots in Germany http://www.transit-port.net/Lists/Robotics.Org.in.Germany.html

Robots in Wikipedia http://en.wikipedia.org/wiki/Robot
    http://he.wikipedia.org/wiki/%D7%A8%D7%95%D7%91% D7%95%D7%98

Robots that jump http://www.plyojump.com/weblog/index.html

Sony's Qrio http://en.wikipedia.org/wiki/Qrio

Sony's QRIO http://www.plyojump.com/qrio.html

Space Robots http://www.robotmatrix.org/space-robot.htm

The Human Element: Robots Everywhere http://www.pcworld.com/article/id,117096-page,1/article.html

Toyota's Humanoid robots http://www.toyota.co.jp/en/special/robot/

Toyota's robots http://www.plyojump.com/toyota.html

Wakamaru robot http://www.mhi.co.jp/kobe/wakamaru/english/

Waseda University, piped robots http://www.uc3m.es/uc3m/dpto/IN/dpin04/2historyindgcr.html

World's greatest android projects http://www.androidworld.com/prod01.htm

Wow-Wee Robosapien http://www.wowwee.com/robosapien/more.html

ZMP's robots http://www.plyojump.com/zmp.html

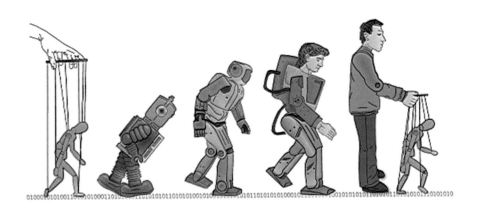

Chapter 3
How to Make a Humanlike Robot

Making humanlike robots is a complex task that requires not only copying the appearance of humans but also replicating the core of what makes us human, including our thoughts, emotions, and capabilities. Effectively, we need to model human movement and behavior, and even the way we think. This task involves many science and engineering disciplines, including mechanical and electrical engineering, materials science, computer science, artificial intelligence (AI), and control. By developing robots that appear and function similar to humans we understand ourselves better and we make robots that we can better relate to, as though we were relating to fellow humans. Making such robots can benefit from advances in biomimetics – the study of nature and its interpretation in biologically inspired technologies.

Constructing an intelligent humanlike robot requires many types of innovations. It takes special materials that are resilient, lightweight, and multifunctional, i.e., making materials that do many things such as compute, sense, and provide structure all at once. It also requires creating robots that can walk using two legs and, while maintaining high stability, being able to traverse complex terrains that include stairs and other obstacles. Also, such robots need sensors for vision, hearing, tasting, and smelling, as well as sensing touch, pressure, and temperature. In addition, they need to be able to interpret the sensory measurements so that they can perceive and be aware of their surroundings as well as the related hazards and risks. Also essential is a lightweight mobile energy generation and storage source that can provide power over a long period for the operation of the robot.

Constructing such robots requires integration of various other capabilities that make it look and act like a human. Humanlike robots need to have effective control and AI

Y. Bar-Cohen, D. Hanson, *The Coming Robot Revolution*, DOI 10.1007/978-0-387-85349-9_3,
© Springer Science+Business Media, LLC 2009

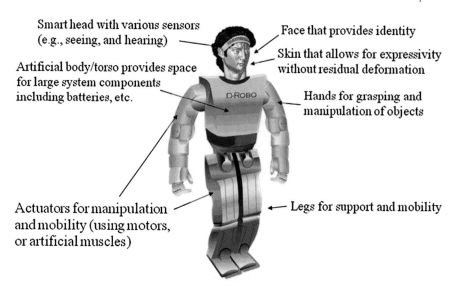

Smart head with various sensors
(e.g., seeing, and hearing)

Face that provides identity

Skin that allows for expressivity
without residual deformation

Artificial body/torso provides space
for large system components
including batteries, etc.

D-ROBO

Hands for grasping and
manipulation of objects

Actuators for manipulation
and mobility (using motors,
or artificial muscles)

Legs for support and mobility

Figure 3.1. Components and functions needed for making humanlike robots.

algorithms to allow them to operate and interact with their environment and with humans.

In a humanlike robot, many parts of the human body and their functions need to be emulated and should operate autonomously. For autonomous operation, the robot needs to be able to conduct self-maintenance, including recharging its batteries before they run out, exploring and learning the environment to minimize its need for help from people, as well as be able to operate safely and avoid situations that can be harmful to humans, itself, and valuable objects in its surrounding. Overall components and requirements for making humanlike robots are illustrated in Figure 3.1 and described in the following sections. In order to distinguish the robotic characteristics, the robot in Figure 3.1 is illustrated as a combination of a human head and a machine body.

HEAD

The head, particularly the face, of the robot is important in that it provides identity to the robot. To look and feel like a living person the head is covered with the equivalent of skin. The skin needs to be highly elastic, to produce facial expressions in a lifelike form. The head needs to have sensors (for hearing, seeing, etc.) to sense the environment from up high and communicate with people in a natural face-to-face conversation. These sensors are needed for safe operation in the interaction with humans and surrounding objects. The equivalent of the brain, which is a computer or multiple microprocessors, does not have to be mounted inside the head, but such placement can minimize unnecessary long wiring to the critical sensors, including the video cameras and sound receivers.

Video cameras provide the computer with the data needed for it to see its environment and obtain cues about the robot's location and surroundings. They also allow viewing and interpreting facial expressions and support the robot's ability to act sociably. Sound sensors provide information about the direction of emitted sound as well as data that the computer can analyze for speech interpretation in a process called "automated speech recognition," which allows for naturalistic verbal communication. For communication with humans, the robot performs voice and facial expressions, and synchronizes lips movements assisted with body gestures. Also, the head can include an artificial nose and tongue to provide information about smell and taste, as occurs in humans.

An important part of making a humanlike robot is the imitation of the human face and its expressions. Our face displays the distinguishing characteristics of our identity and is a critical part of communicating as social and intelligent individuals. Also, the face houses important sensing capabilities that include seeing, hearing, smelling, and tasting. Through facial and verbal expressions we communicate our thoughts to other people. By mimicking natural communication features, robots can express themselves in a way that we can intuitively understand. Such expressivity can turn robots into beings that command our love and admiration. An essential part of making robots that we can relate to is the need to make the face as fluidly humanlike as possible. Some humanlike robots are already making realistic facial expressions that can momentarily be mistaken for a real person while conversing with humans. Future robots must maintain this suspension of disbelief during more complex, useful social exchanges, a task that requires further improvements to both software and the hardware that produces the facial expressions.

ARTIFICIAL SKIN

The use of artificial skin that looks and feels like natural human skin is important to the lifelike appearance of a humanlike robot. In Japan, the cosmetic manufacturer Kao Corporation and a Keio University research team developed artificial skin that feels like human skin. This artificial outer covering consists of 1-cm thick "dermis" of elastic silicone with a 0.2-mm thick "epidermis" of firm urethane. Numerous tiny hexagonal indentations are etched into the urethane epidermis to provide it with a lifelike texture. According to its developers, 10 out of 12 people agreed that it felt like regular human skin. The similarity of this artificial skin to the skin of natural humans was also confirmed by test machines.

One recently developed material for use as an artificial skin is called Frubber. This material was found effective for making facial expressions, since it requires minimal force and power to produce natural-looking large deformations. Using this material, several heads that make facial expression have been produced (see, for example, Figure 3.2 below). The head that is shown in Figure 3.2 serves as a platform for engineers worldwide who are developing artificial muscles (see later in this chapter) and need a robotic platform to test their developed actuators. This head is currently located at

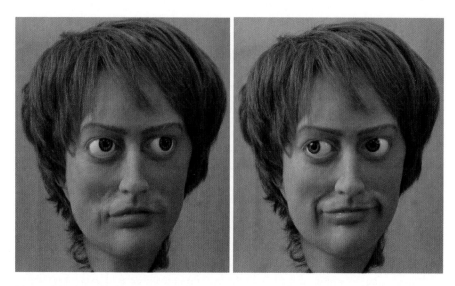

Figure 3.2. Using Frubber skin, a humanlike head that makes facial expressions was produced as a platform for testing artificial muscles.

the NonDestructive Evaluation and Advanced Actuators lab at the Jet Propulsion Laboratory (JPL) in Pasadena, California.

If an artificial skin can be made modifiable and controllable, it may become possible for a doctor to sense the condition of the skin of a human patient located at a remote location. Such conditions may involve breast exams or other diagnostic procedures that require feeling where the real person is at a distant location while the doctor uses a haptic interface to feel with his/her own hands. Other applications may include huggable toys that children would find enjoyable to play with.

EXPRESSIVE FACE

Humans are quite sensitive to natural expressivity. Facial expressions are primarily nonverbal communication means of conveying social information among humans. They are generated by one or more motions or positions of the face muscles. In an effort to quantify the characteristics of facial expressions various researchers categorized and grouped the expressions. A simplistic graphic illustration of the basic facial expressions is shown in Figure 3.3. The X-axis represents the level of excitement from low to high, while the Y-axis shows the level of positivity from negative feelings to positive feelings. Accordingly, the quadrants are as follows: the top right shows the face of a happy or praised person; the top left shows the face of a comfortable/soothed person; the bottom left shows the face of a sad person; and the bottom right shows the face of an angry or astonished person.

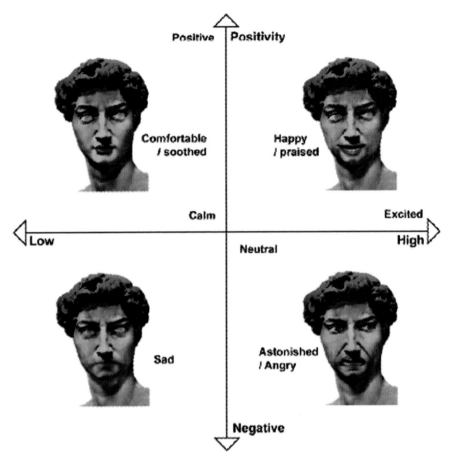

Figure 3.3. Graphic illustration of facial expression categories.

These facial expressions are relatively easy to tell apart, but there are more complex expressions that may not be easy to identify or distinguish from others. Because the human face has only a limited range of movement, expressions rely on relatively small differences in the proportion and relative position of facial features. Reading the expression sometimes requires considerable sensitivity. Further, some human faces are often falsely read as expressing certain emotions, even though the expression is intended to be neutral.

Realistic mimicking of the human face is a great challenge, and any imperfection stands out to human observers as a fault. Effective expression is equivalent to bandwidth in communications because it provides the ability to communicate in much more detail. Humans are visual creatures, and our ability to recognize faces and facial expressions is hard-coded within the human nervous system. For this reason, mimicking the human facial appearance, shape, and movements is an especially challenging task where the developer needs to be able to generate lifelike expressions that include wrinkles and folds that are dormant in the nonexpressive position of the face. Movements on the opposite sides of the face need to be generated on the skin in different forms of folding and grouping. Further, the human skin is affected by the muscle pulling in other areas through internal tissue layers that are characterized by various physical properties that are inherent to the complex biological media of the natural human face. These characteristics are very difficult to emulate by simply using standard rubber or silicone, since artificial skins respond differently to the action of motors than does the human face to the action of natural muscle fibers. When making simulations of facial expressions the actuation associated with the related movement requires attention to the various facial components to achieve all the complex range of natural expressions. When a robot speaks, one of the critical aspects of making its expression realistic and lifelike is the synchronization of its speech with the lips movement, facial expressions, and body gestures.

Using his Frubber material, the coauthor and his company, Hanson Robotics Inc., produced an expressive humanlike head that replicated Albert Einstein. The head consumes <10 W at 6 V, allowing it to operate for hours using a small set of rechargeable batteries. Thus, the movement of this skin for facial expressions requires much lower power compared to the skins that were used to produce the other robots described in Chapter 2. This low-power requirement benefited the constructed robot, where the head was mounted onto the walking Hubo robot body (made by KAIST, S. Korea; see Chapter 2). The integrated Einstein-like head with a humanoid robot body was called "Albert Hubo," and it was developed in celebration of the 100-year anniversary of the theory of relativity. On November 19, 2005, this robot was demonstrated at the APEC summit that was held in Busan, South Korea, where it greeted and shook hands with numerous world leaders, including the president of the United States, George W. Bush.

TORSO AND POWER SOURCES

Since a robot does not need lungs or a digestive system, at least not yet, the chest can be filled with various system components that take up space, including rechargeable batteries and other energy-producing mechanisms. Although not sufficiently mature yet, there are efforts to develop the equivalence of a digestive system, where food is processed to create power using electrochemical conversion techniques. Alternatively, miniature fuel cells are increasingly being used. Dynamic power technologies, such as micro-turbines and miniaturized internal combustion engines, are also being explored. These mechanisms are still far from being usable in commercial humanlike robots. One of the concerns regarding the implementation of these energy generation alternatives is the noise that they create.

An attractive solution to maximizing the power-to-weight ratio is to develop structurally embedded energy storage components that are compatible with construction materials and the related processing techniques. These storage components need to be integrated with energy harvesting mechanisms/devices that may include solar cells and thermoelectric converters. High energy power storage/sources based on lithium chemistry are already available and are providing miniature rechargeable sources of electric power. Such batteries are shaped in tubular and flat configurations, allowing their multifunctional use as energy storage devices and as parts of the robot construction components. This capability of using batteries as structural members is a characteristic of biological systems and provides benefits in lower mass and more efficient operation.

ARMS AND LEGS

Similar to the action of natural appendages, the mobility and manipulation of objects are done by the hands and legs of the robot. Further, having the ability to walk on two legs and maintain dynamic stability is now an established capability of many biped humanoids and humanlike robots.

For efficient use of power while walking the ability to cycle energy through the springy action of natural musculature is important. Currently, in most of the reported humanlike robots (see examples in Chapter 2), electric motors are operating the joints of the fingers, wrists, ankles, elbows, and many other parts of the robot's body. To act in a natural way, sensors are mounted on the legs, arms, and hands, including pressure sensors to determine the grip level and touch sensors to interpret tactile impressions. Sensors are used to trigger needed action if the robot is exposed to excessively low or high temperatures or suffers overheating of any of its parts, thus assuring safe operation. The data from the sensors are communicated to a central processing computer that may reside in the head or the chest, but the robot can have a distributed decision-making capability. Making decisions locally is analogous to biological systems, and it allows for high-speed response similar to our reflexes.

Bipedal Ambulation

Several decades ago toy makers were already able to create machines that were capable of walking. In the 1960s, a kinematic mock-up of a pair of legs was constructed in Petersburg, Russia, by Kemurdjian and his co-investigators. In the early 1980s, Mark Raibert studied legged mobility at his Leg Laboratory at Carnegie Mellon University. In 1986, he moved to the Massachusetts Institute Technology (MIT), where he continued this effort, focusing on stability using dynamic balance. He soon produced a legged mechanism that he called the Raibert Hopper with which he could study dynamic balance. His mechanism kept itself stable upright in steady state by controlling the required next step. He followed the one-legged Raibert Hopper with a 3-D bipedal mechanism.

Another pioneer in the study of legged mobility is Tad McGeer, who modeled machine walking in an effort to emulate human mobility using recycled energy and minimal power. He suggested a passive mechanism using self-stabilizing natural dynamics. Other researchers studied the stability of passive walking with leg-swing motion and demonstrated a mechanism of flexion and extension of the knee joint of the swing leg. This study was aimed towards realizing a robust passive walking robot that could walk stably on irregular terrain.

Generally, bipedal humanlike robots rely on static or quasi-static stability that controls the walking process and prevents fall. The walking speed is maintained such that with each footstep the robot catches itself from falling. This technique of walking is not effective for dynamic running or traversing of complex terrain. As opposed to the use of springy gaits for walking, which is seen in animals, most walking robots rely on rigid mechanical systems and therefore do not have shock absorption, which is the advantage of spring-dampener stability (called "preflex") or the inverse-pendulum energy recycling.

HANDS AND MANIPULATION

An important part of the functionality of humanlike robots is the ability to use hands for object grabbing and manipulation. Developing these functions of the hands can be quite challenging when the robot is operating in unpredictable environments and when possibly the desired object is moving. This requires the ability to combine visual information and tactile object recognition with dynamically responsive robotic capabilities. For this purpose, robots need to operate with hands similar to the way humans do. The manipulation device must be guided to the object using a control system with input from machine vision having a high-speed and high-resolution imaging capability. Then, the mechanical system needs to reach the target objects with sufficient flexibility and dexterity. Also, the robot needs to contain enough sensors and the right control software to apply just the right amount of force necessary to lift the object without system failure or damage to the object. Today, artificial hands are widely used in robots and in prosthetics with great similarity of appearance and performance as natural hands; examples are shown in Chapter 4.

ACTUATORS AND ARTIFICIAL MUSCLES

Actuators are used as emulators of muscles and are responsible for the movement and mobility of robots and their appendages. The types of actuators that are generally used include: electric, pneumatic, hydraulic, piezoelectric, shape memory alloys, or ultrasonic motors (USM). Motors are widely used to perform the movements of humanlike robots, but they behave differently than our natural muscles and have a totally different operation mechanism. Specifically, electromagnetic motors (DC and AC types) are used with gears to compromise the speed of rotation to obtain a higher torque. Both, hydraulic and electric motors exhibit very rigid behavior and they require relatively complex feedback control strategies to enable them to have compliant performance. Generally, hydraulic and pneumatic motors are operating at low speeds. They require a compressor, which is bulky, heavy, expensive, and acoustically noisy, and therefore not highly attractive. For this reason, electric motors are currently the key actuators that are used for mobility and manipulation in humanoids and humanlike robots.

Piezoelectric motors, particularly in a stack configuration, are a form of electric motor used to generate micrometer-size movements with high force, but these require application of high voltage. One of their major advantages is their ability to provide extremely high positioning accuracy and self-sensing capability. Also, they can be used in static or dynamic applications for generating and handling high forces. In order to excite large movement ultrasonic motors are driven by piezoelectric disks that excite elastic waves and move objects that are pressed onto their activated surface. The wave provides a propelling action that moves the rotor or slider (depending on the type of movement that is produced, i.e., rotary or linear, respectively), which is the object that is pressed onto the surface of the stator. Ultrasonic motors are driven mostly at frequencies of about 20 kHz, and to maximize their operation they are used in resonance. The speed and torque of the rotary motors depends on their diameter, where speeds of several tens of revolutions per minute and torque of several lb-in are generated by such motors with a 2.5–5 cm (1–2 in.) diameter.

Natural muscles are both compliant (flexible or rubbery) and exhibit linear behavior. These capabilities are important to emulate since they address key control requirements for robotic applications with lifelike performance. The closest to have the potential to emulate natural muscles are the electroactive polymers (EAP) that have emerged in recent years and gained the name "artificial muscles." In the 1990s, the developed materials were very weak in terms of the ability to move or lift objects. Recognizing the need for international cooperation among the developers, users, and potential sponsors, in March 1999 the first annual international EAP Actuators and Devices (EAPAD) Conference was organized. This conference was held by the technical society SPIE as part of its annual Smart Structures and Materials Symposium. At the opening of the first conference, Yoseph Bar-Cohen posed a challenge to researchers and engineers worldwide to develop a robotic arm actuated by artificial muscles that could win an arm wrestling match against a human opponent. See Figure 3.4, which illustrates the wrestling of a human arm with a robotic one driven by artificial muscles.

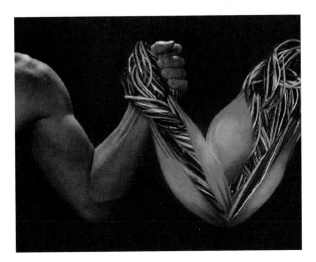

Figure 3.4. Arm wrestling challenge for artificial muscles against human.

In 2005, three groups of scientists and engineering students were ready with EAP-actuated robotic arms for the first arm wrestling match against a human. The contest was held on March 7, 2005, at San Diego, California, as part of the annual SPIE's EAPAD Conference, and it was organized with assistance from the U.S. organization ArmSports, which provided the wrestling table and a representative as one of the contest judges. The human opponent was a 17-year-old straight-A high school female student from San Diego. In Figure 3.5, a photo shows a member of the group of students from Virginia Tech preparing for one of the three wrestling matches that were held. Even though the female student, whose name was Panna Felsen, won in all the three wrestling matches, this was a very important milestone in the field.

Since the currently developed robotic arms with EAP actuators are not at a level that allows winning against humans, Bar-Cohen changed the focus of the near future contests. As of 2006, the contest was altered to measuring arm capability and comparing the data of the competing arms. A device was created jointly by the UCLA professor, Qibing Pei, and his students from UCLA as well as Bar-Cohen's group at JPL (see Figure 3.6) that measured the speed and pulling force capability of the arms. The test

Figure 3.5. The robotic arm driven by artificial muscles, made by Virginia Tech students, is being prepared for the match against the human opponent, Panna Felsen, a 17-year-old high school student from San Diego.

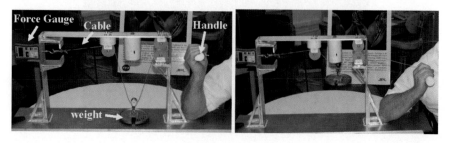

Figure 3.6. The device that is used to test the force and speed of the EAP-actuated robotic arms.

device was equipped with a pulling cable that was connected to a force gauge. To gauge the speed a 0.5-kg weight was mounted on the cable, which had to be lifted to the top of the device, and the time it took to reach the top was measured. To establish human reference data, Panna Felsen was invited again to the event, and her capability was measured. The 2006 results of three participating arms have shown about two orders of magnitude slower and weaker force capability than the student. In a future conference, once improvements in developing such arms reach sufficiently high level, a professional wrestler will be invited for the next human/machine wrestling match.

SENSORS AS ARTIFICIAL SENSES

Our senses are critical to our existence as living creatures, and they are emulated using sensors. Natural senses provide input to the central nervous system about the environment around it and within the body, and our muscles are instructed to act based on our brain's interpretation of the received sensory information. Biological sensory systems are extremely sensitive and limited only by quantum effects.

The key sensors in the head are the eyes, which in robots are emulated using cameras, and the ears, which are emulated by acoustical detectors, including microphones. Other sensors that are being developed include an artificial nose for providing the equivalence of smelling and an artificial tongue for emulating tasting.

Generally sensors provide electrical signal data about selected parameters that are converted using a physical or mechanical effect that is measured or monitored. In robots there are various sensors that are used; some of them are described below, while hearing and seeing sensors are described later.

Proprioceptive sensors provide insights into the robots' own conditions, including position, orientation, and speed of movement of body and joints. In humans, balance and stability are sensed by the inner ears; in robots, the sense and control of balance is provided using gyros, accelerometers, and force sensors. *Accelerometers* are used to determine the acceleration along a particular axis, and the speed of movement is calculated by integration over time. *Tilt sensors* are used to measure the inclination. *Force sensors* are used to measure contact forces and are placed in the hands and feet of the robot to measure the contact with the environment. Gyrometers measure in three dimensional coordinates the rate of angular rotation along a given axis. *Position sensors* are used to indicate the physical position of the robot's joints and can provide velocity data by derivation over time. Further, *speed sensors* are also used to provide information about the position of the robot's joints.

Exteroceptive sensors provide the robot with information about its surrounding environment, namely the world around the robot. These sensors are important in allowing the robot to interact with the world in real-time and realistic terms. Exteroceptive sensors are used for external perception and include *proximity sensors,* which measure the relative distance between the robot and objects in its environment. These sensors may include stereo *cameras, laser ranging, sonar, infrared* or *tactile sensors* (including bumps, whiskers, or feelers as well as capacitive or piezoresistive sensors). Sensors are also used to measure forces and torques that are transmitted between the robot and objects that are in contact with it.

Artificial Nose

The sense of smell is our means of analyzing chemicals by sampling airborne molecules. This allows us to determine the presence of hazardous gases as well as helping us to find and enjoy good food and the benefits of other pleasant odors. The sense of smell alerts us to such dangers as smoke from fire and spoiled food. It is estimated that our nose can distinguish between up to 10,000 different smells.

Imitating the sensing capability of the human nose has important applications, and efforts to make such a sensor have been under development since at least the 1980s. Several devices have been built and tested that emulate the smelling ability of the nose using chemical sensor arrays. These arrays are weakly specific chemical sensors, controlled and analyzed electronically, mimicking the action of the mammalian nose by recognizing patterns of response to vapors. Unlike most existing chemical sensors, which are designed to detect specific chemical compounds, the sensors in an electronic nose are not specific to any one compound, but have overlapping responses. Gases and their mixtures can be identified by the pattern of the responses of the sensors in the array. The technology is not small enough, cheap enough, or mass producible enough yet for widespread use in humanlike robots, however.

Artificial Tongue

The sense of taste is another chemical analyzer in our body. This sense examines dissolved molecules and ions using clusters of receptor cells in the taste buds. Generally, there are five primary taste sensations, including salty, sour, sweet, bitter, and umami (savory). Similar to the electronic nose, researchers are developing electronic tongues to mimic biological sensory capabilities. Electronic tongues are used in the analysis and recognition (classification) of liquids and they consist of arrays of sensors that are capable of detecting or sensing many types of materials rather than being tailored to specific ones. Also, their system consists of data acquisition elements, and analytical tools. This technology overcomes the limitations of human sensing, including individual variability, inability to conduct online monitoring, subjectivity, adaptation, infections, harmful exposure to hazardous compounds, and effect of mental state. E-tongues are increasingly being used in such applications as monitoring food taste and quality, noninvasive diagnostics, searching for chemical/biological weapons, drugs, and explosives, as well as environmental pollution monitoring. Of course, these sensors need to be small and portable, and ideally inexpensive, before they can be widely used in humanlike robots.

ARTIFICIAL VISION

The ability to create and process high-quality digitial images can be attributed to the current powerful capabilities of computer imaging hardware and software. Algorithms that allow recognition of objects and faces are quite effective today and are increasingly improving in performance and speed. Such algorithms are operational via personal computers, or using a small microprocessor, which can recognize faces, facial expressions, gestures, obstacles, objects, and even text. These capabilities are critical for making robots autonomous. The use of computer vision has been successfully applied by the JPL in various NASA missions, including the two Mars Exploration Rovers (Spirit and Opportunity) that were launched in 2003. Other beneficiaries of the developed capabilities are the military, home and office security, medicine, and entertainment. The use of vision interpretation algorithms allows robots to navigate in unknown terrain while avoiding obstacles as well as assisting in verbal communications in conversations with humans.

In one example of the numerous ambitious computer vision projects around the world, an interdisciplinary research team consisting of computer scientists at Carnegie Mellon University and psychologists at University of Pittsburgh was formed to develop software for rigorous, quantitative analysis of facial expression in diverse applications. This team developed effective software, which is known as the Automated Facial Image Analysis (AFA) system, and it is capable of automatically recognizing facial action and analyzing facial behavior in real time. The system can deal with spontaneous facial behavior in people of widely varying ages (including infants) and ethnic backgrounds. This work has opened areas of research in clinical, developmental, and mental health science, and it generates knowledge about the dynamics of individual and interpersonal behavior. One difficulty of the AFA system is that it is not an effective tool for real-time image analysis. For automated identification of facial expressions in real time, a recent study at the University of California at San Diego's Machine Perception Lab led to an algorithm that is applicable for use in humanoids and humanlike robots.

The success of the algorithms of facial identification and recognition has been impressive, and commercial applications are becoming widespread. Artificial vision is used in security systems allowing photo identification of individuals in the system database. Following the events related to September 11, 2001, efforts were made to include such image recognition systems in airports across the United States, but the level of false positives is still too high for implementation.

ARTIFICIAL INTELLIGENCE

Artificial intelligence provides smart control to the operation and functionality of robots. Artificial Intelligence is the branch of computer science that studies the computational requirements for such tasks as perception, reasoning, and learning, and to allow the development of systems that perform these capabilities. The field seeks to advance the understanding of human cognition, understand the requirements for intelligence in general, and develop artifacts such as intelligent devices, autonomous agents, and systems that cooperate with humans to enhance their abilities. AI researchers are using models that are inspired by the computational ability of the brain and explaining these abilities in terms of higher-level psychological constructs such as plans and goals.

Progress in the field of AI has enabled better scientific understanding of the mechanisms underlying thought and intelligent behavior and their embodiment in robots. The field of AI is providing important tools for humanlike robots, including knowledge capture, representation and reasoning, reasoning under uncertainty, planning, vision, face and feature tracking, language processing, mapping and navigation, natural language processing (NLP), and machine learning. AI-based algorithms are used where case-based and fuzzy reasoning are combined to allow for automatic and autonomous operation. Even though AI led to enormous successes in making smart computer controlled systems, still the capability is far from resembling the human intelligence.

To best achieve humanlike behavior, one must achieve humanlike cognition. But this is a daunting task, as the full complexity of human cognition is not well understood, and in fact most neuroscientists say that we have barely scratched the surface of understanding the human mind. In the meantime, however, we can create robots that are useful and entertaining, by constructing control systems that use the best of today's AI, machine perception, and motion control software, to imitate human behaviors for specific applications. If such a robot needs to perform as a nurse, then the expertise of a nurse can be painstakingly catalogued and represented as a series of rules, known as an expert system, which then can guide a robotic nurse through the steps of problem-solving to achieve the needed health care for a patient. Since the 1970s, such expert systems have been routinely employed in many industries, including health care. However, if more flexible behavior is required (for example, a patient presents a case outside the Nursebot's expert rules), the robot needs to have some creative problem-solving capabilities, which is currently the topic of the most ambitious AI research in the world.

To pave a path of useful transition toward truly flexible humanlike behavior, scientists are combining scripted expert-type robots with more open-ended flexible AI processes. Although the resulting software may not resemble the structure of the biological human mind, it allows a robot to recover intelligently in unexpected situations and to learn from the results. This is essentially the architecture that Marvin Minsky describes in his book *The Emotion Machine* (Minsky, 2006). His theories need to be implemented in a practical way, which will take many years to produce. In the meantime, as today's useful humanlike robots prove by their behavior, we do not need to implement a complete humanlike behavior in order to have delightful and beneficial humanlike robots.

In another area, speaking via voice synthesizers, quickly doing language translations, and processing/recognizing speech are several of the important benefits today of using AI. Such verbal communication capabilities are routinely used to respond to phone calls, as with an automatic operator, and we have already become accustomed to talking to such machines rather than humans when we call any service provider today. This includes your bank, airline, credit card companies, telephone operators, and many others. The performance of these systems continues to improve, especially thanks to recent boosts in computing capability that is known as NLP. Effective integration of the techniques to synthesize sound and to recognize speech with synchronized facial expression has enormous value to making humanlike robots more lifelike. Also, it can provide a great user interface in various digital technologies that use verbal expressions. The more realistically the expressions can be performed, the more effectively it grasps and holds the attention of the audience, and the better it focuses our attention on the content of the expressed message rather than on the mechanism that generates it. To make speech sound realistic the operating system must coordinate the sound, lips, and facial movements, as well as the hand gestures and the eyes blinking while periodically looking at the audience or the person with whom the conversation is being conducted. Also, while interpreting human speech the robot needs to use its two eyes (i.e., cameras that sends video streams for image analysis) to see the person and show attentiveness. In doing so, the head needs to track the face and perceive the facial expressions of the person who is being listened to.

To understand speech, the microprocessor turns audio-signals received from the microphones into digital wave files, and then converts these files into words and sentences via automatic speech recognition (ASR). The meaning of the obtained text is analyzed and interpreted via various NLP techniques. Based on the resulting "understanding" of the speech content the computer decides how to respond. Usually, for a humanoid and humanlike robots, the response is either spoken and gestural or both. For simulating spoken responses, the computer uses a synthesizer that converts the text to speech (TTS) emulating human sound. Progress in this area has been remarkable, but the use of the synthesizer is still quite limited, as indicated from the number of words (mostly 700 words or less) that the reported robots are able to recognize (see Chapter 2). The benefits of this technology would be enormous in relation to commerce between countries (i.e., international customer service support, import and export, etc.), tourism, and military applications.

Under a recent DARPA contract efforts are being made to develop a translation system and to install the capability in palm size devices for use in communication with local people in various countries. An effective open source speech recognition engine is the Sphinx, which was developed by Carnegie Mellon University under DARPA funding. When given a large volume of words the Sphinx was shown to have recognition capability at a success rate of over 50% and as high as 70%. The CMU Robust Speech Recognition group that developed the Sphinx is currently working to further improve the accuracy of their speech recognition systems in order to allow operating in challenging acoustical environments that include noise and interferences. Just as in human interpretation of speech, it is not sufficient to convert the heard words and sentences to understandable content. There is also a need to use other inputs that include the context of the conversation, the social environment, the facial expressions, lips-speech synchronization, and gestures. The use of such inputs combined with speech recognition is still a great challenge for machines.

SUMMARY

Humanlike robots are complex machines that require advancements in many disciplines to effectively mimic the appearance and functionality of humans. Some of the required capabilities include the emulation of the movement and behavior of people, and simulation of human thought to allow flexible and intelligent autonomous operation. This grand challenge of making humanlike robots is a multidisciplinary task that spans over numerous science and engineering fields. Although progress has been rapid, many obstacles in AI and cognitive science must be overcome to achieve flexible thinking robots, the kind that realize the science fiction dream of robots that are the peers of humanity. In the meantime, we will continue to build robots that look and behave just enough like us to astound us, entertain us, tutor and train us, and serve us in other intermediate applications. Although these robots may not be as smart or lifelike as people, they evoke those questions within us that we encounter in science fiction: What does it mean to be human? Are we humans simply no more than extremely complex machines? And will our technology stop progressing once it creates machines that are humanlike in intelligence, or will the trend continue, resulting in machines that are far more intelligent than we are, thus accelerating the evolution of the human species forward in an artificial way?

BIBLIOGRAPHY

Books and Articles

Altarriba, J., D. M., Basnight, and T. M. Canary, "Emotion representation and perception across cultures", in W. J. Lonner, D. L. Dinnel, S. A. Hayes, and D. N. Sattler (Eds.), *Online Readings in Psychology and Culture* (Unit 4, Chapter 5), www.wwu.edu/~culture, Center for Cross-Cultural Research, Western Washington University, Bellingham, Washington (2003).

Bar-Cohen, Y. (Ed.), *Electroactive Polymer (EAP) Actuators as Artificial Muscles – Reality, Potential and Challenges*, 2nd Edition, SPIE Press, Bellingham, Washington, Vol. PM136, (March, 2004).

Bar-Cohen, Y., (Ed.), *Biomimetics – Biologically Inspired Technologies*, CRC Press, Boca Raton, FL, (November, 2005).

Bar-Cohen, Y., and C. Breazeal (Eds.), *Biologically-Inspired Intelligent Robots*, SPIE Press, Bellingham, Washington, Vol. PM122, (May, 2003).

Bar-Cohen, Y., X. Bao, and W. Grandia, "Rotary Ultrasonic Motors Actuated By Traveling Flexural Waves," *Proceedings of the SPIE International Smart Materials and Structures Conference*, SPIE Paper No. 3329-82 San Diego, CA, (March 1–6, 1998).

Bartlett, M.S., Littlewort, G.C., Lainscsek, C., Fasel, I., Frank, M.G., Movellan, J.R. Fully "Automatic facial action recognition in spontaneous behavior" *7th International Conference on Automatic Face and Gesture Recognition*, (2006), pp. 223–228.

Breazeal, C., *Designing Sociable Robots*. MIT Press, Cambridge, MA (2002).

Dunn, E., and R. Howe, "Foot placement and velocity control in smooth bipedal walking," *Proceedings of the IEEE International Conference on Robotics and Automation*, (1996) pp. 578–583.

Full, R. J., and K. Meijir, "Metrics of Natural Muscle Function," Chapter 3 in Bar-Cohen Y. (Ed.), *Electroactive Polymer (EAP) Actuators as Artificial Muscles - Reality, Potential and Challenges*, 2nd Edition, SPIE Press, Bellingham, Washington, Vol. PM136, (March, 2004), pp. 73–89.

Hans, M., "The Control Architecture of Care-O-bot II." In: Prassler, E.; Lawitzky, G.; Stopp, A.; Grunwald, G.; Hägele, M.; Dillmann, R.; Iossifidis, I. (Eds.): *Advances in Human-Robot Interaction. Book Series of Springer Tracts in Advanced Robotics*, Vol. 14 (2004), pp. 321–330.

Hanson, D., *Humanizing interfaces – an integrative analysis of the aesthetics of humanlike robots*, PhD Dissertation, The University of Texas at Dallas (May, 2006).

Hardenberg, H. O., *The Middle Ages of the Internal combustion Engine*, Society of Automotive Engineers (SAE), (1999).

Hirai, K., M. Hirose, Y. Haikawa, and T. Takenaka, "The development of Honda humanoid robot," *Proceedings of IEEE International Conference on Robotics and Automation*, (1998).

Ikemata, Y., K. Yasuhara, A. Sano, and H. Fujimoto, "A Study of the Leg-swing Motion of Passive Walking," proceedings of the 2008 IEEE International Conference on Robotics and Automation (ICRA), Pasadena, CA, USA, (May 19–23, 2008), pp. 1588–1593.

Kajita, S., and K. Tani, "Experimental study of biped dynamic walking in the linear inverted pendulum mode," *Proceedings of the IEEE International Conference on Robotics and Automation*, (1995).

Kazerooni, H., and Guo J., "Human Extenders", *Transactions of the ASME, Journal of Dynamic Systems, Measurements, and Control*, Vol. 115, (1993) pp. 281–290.

Kurzweil, R., *The Age of Spiritual Machines: When Computers Exceed Human Intelligence*, Penguin Press, (1999).

Lamere, P., P. Kwok, W. Walker, E. Gouvea, R. Singh, B. Raj, and P. Wolf, "Design of the CMU Sphinx-4 decoder," *Proceedings of the 8th European Conference on Speech Communication and Technology*, Geneve, Switzerland, (September, 2003), pp. 1181–1184.

Larminie J., *Fuel Cell Systems*, 2nd Edition, SAE International, (May, 2003).

McGeer, T., "Passive dynamic walking," *International Journal of Robotics Research*, Vol. 9, (1990) pp. 62–82.

Minsky, M. *The Emotion Machine,* Simon & Schuster (November 7, 2006)

Miura, H., and I. Shimoyama, "Dynamic walk of a biped," *International Journal of Robotics Research*, Vol. 3, No. 2, (1984) pp. 60–74.

Mizuuchi, I., T. Yoshikai, Y. Nakanishi, Y. Sodeyama, T. Yamamoto, A. Miyadera, T. Niemelä, M. Hayashi, J. Urata, and M. Inaba, "Development of Muscle-Driven Flexible-Spine Humanoids," *Proceedings of Humanoids 2005, part of the 5th IEEE-RAS International Conference on Humanoid Robots*, (December, 2005), pp.339–344.

Moriyama, T., J. Xiao, J. Cohn, and T. Kanade, "Meticulously detailed eye model and its application to analysis of facial image," *IEEE Transactions on Pattern Analysis and Machine Intelligence*, Vol. 28, No. 5, (May, 2006), pp. 738–752.

Plantec, P. M., and R. Kurzwell (Foreword), *Virtual Humans: A Build-It-Yourself Kit, Complete With Software and Step-By-Step Instructions*, AMACOM/American Management Association; (2003).

Pratt, G. A., "Legged Robots at MIT: What's new since Raibert," *Research Perspectives, IEEE Robotics and Automation Magazine*, (September, 2000), pp. 15–19.

Raibert, M., *Legged Robots that Balance*, Cambridge, MA: MIT Press, (1986).

Sherrit, S., Y. Bar-Cohen, and X. Bao, "Ultrasonic Materials, Actuators and Motors (USM)," Section 5.2, Chapter 5 "Actuators," Y. Bar-Cohen (Ed.) *Automation, Miniature Robotics and Sensors for Nondestructive Evaluation and Testing*, Volume 4 of the Topics on NDE (TONE) Series, American Society for Nondestructive Testing, Columbus, OH, (2000), pp. 215–228

Twomey, K., and K. Murphy "Investigation into the packaging and operation of an electronic tongue sensor for industrial applications," *Sensor Review, Emerald Group Publishing Limited*, Vol. 26, Issue 3, (2006) pp. 218–226.

Uchino, K., *Piezoelectric Actuators and Ultrasonic Motors*, Kluwer Academic, (1996)

Ueha, S., Y. Tomikawa, and M. Kurosawa, *Ultrasonic Motors: Theory and Applications*, Clarendon Press, (1993)

von der Malsberg C., and W. Schneider, "A neural cocktail party processor," *Biological Cybernetics,* Vol. 54, (1986), pp. 29–40.

Westervelt, E. R., J. W. Grizzle, C. Chevallereau, J. H. Choi, B. Morris, *Feedback Control of Dynamic Bipedal Robot Locomotion (Control and Automation)*, CRC Press, Boca Raton, FL, (June 26, 2007).

Yamaguchi, J., E. Soga, S. Inoue, and A. Takanishi, "Development of a bipedal humanoid robot – Control method of whole body cooperative dynamic biped, walking," *Proceedings of the IEEE International Conference on Robotics and Automation*, (1999), pp. 368–374.

Internet Websites

Actoid DER http://www.kokoro-dreams.co.jp/english/index.html

Android world http://www.androidworld.com/prod01.htm

Articles in Electronic Design http://www.elecdesign.com/Articles/Print.cfm?ArticleID=14763 http://www.elecdesign.com/Articles/ArticleID/12830/12830.html

Carnegie Mellon contributions to walking robots http://www.newdesign.cs.cmu.edu/archive/2007/pages/RHO FReleasePhotos.pdf

Female Guide Robot http://www.plasticbamboo.com/2006/10/05/female-guide-robot-from-sanrio/

Game machines play (including World Cup) http://www.pbs.org/saf/1208/resources/resources-1.htm

Gynoid (Female-like robots) http://en.wikipedia.org/wiki/Gynoid

History of walking machines http://agrosy.informatik.uni-kl.de/wmc/historical.php
    http://www.humansversusandroids.com/androids/real.htm]
Honda's Asimo http://asimo.honda.com/
Humanlike robots http://www.flashedblogs.com/filedunder/robots/193.html?page=6
Humanoid robots http://en.wikipedia.org/wiki/Humanoid_robot
Humanoid robots http://www.droidlogic.com/
Kotaro http://www.jsk.t.u-tokyo.ac.jp/~ikuo/kotaro/index-e.html
Legged Robots at MIT
    http://ieeexplore.ieee.org/iel5/100/18981/00876907.pdf?arnumber=876907
NEC's PaPeRo Robot http://www.incx.nec.co.jp/robot/english/intro/intro_01.html
RoboCup http://www.robocup.org/
RoboGames (formerly Robolympics) http://en.wikipedia.org/wiki/Robolympics
Robosapien http://www.robotbooks.com/robosapien.htm
Robot books http://www.robotbooks.com/
Robotic faces http://www.androidworld.com/prod04.htm
Robots that jump http://www.plyojump.com/weblog/index.html
Sony's QRIO http://www.plyojump.com/qrio.html
Space Robots http://www.robotmatrix.org/space-robot.htm
Standard software platform http://www.roboticstrends.com/displayarticle1022.html
Talking heads http://www.cse.unr.edu/~rakhi/th_avsp99.pdf
Toyota's Humanoid robots http://www.toyota.co.jp/en/special/robot/
Toyota's robots http://www.plyojump.com/toyota.html
Wakamaru robot http://www.mhi.co.jp/kobe/wakamaru/english/
Waseda University, piped robots http://www.uc3m.es/uc3m/dpto/IN/dpin04/2historyindgcr.html
Wow-Wee Robosapien http://www.wowwee.com/robosapien/more.html
ZMP's robots http://www.plyojump.com/zmp.html

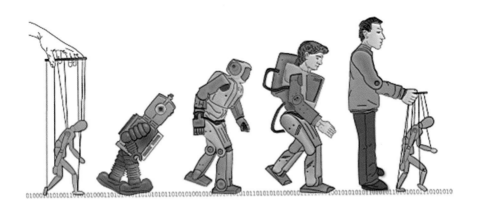

Chapter 4
Prosthetics, Exoskeletons, and Bipedal Ambulators

From the dawn of humanity the inventive use of tools has distinguished us from animals. This trend has accelerated as the number of tools has grown, which has led to even better tools. Writing led to engineering; industrial tools led to automation; and computers led to artificial intelligence. For centuries, our tool-making technologies have also subtly altered our human abilities by augmenting our physical capabilities; extending our lives; helping us communicate and preserve our ideas; and even helping us to solve problems that our unaided minds could not. More profoundly still, our tool making has begun to alter human nature itself, especially with the interaction of the biosciences and new medical technologies.

As modern computing has helped us unlock mysteries of the physics and chemistry of human genes and proteins, it has also addressed the mysteries of the human nervous system. Thanks largely to the power of today's computers, combined with advanced brain scanning devices and software such as functional Magnetic Resonance Imaging, we now understand human cognition better than ever and are further discovering mechanisms of thought at an unprecedented pace. The emerging discipline of computational neuroscience takes the related theories and represents them as mathematical models and software algorithms, using them to test the theories and also to generate rapid improvements in artificial intelligence. These understandings are resulting in machines that think more like humans.

It is interesting to note that not only machines are becoming more humanlike in thoughts but also humans are incorporating computers directly into their brains. In the last few decades, technologies have been developed that directly interface with

Y. Bar-Cohen, D. Hanson, *The Coming Robot Revolution*, DOI 10.1007/978-0-387-85349-9_4,
© Springer Science+Business Media, LLC 2009

the human nervous systems. These include cochlear implants for the profoundly deaf, artificial retinas for blind people, and neurally controlled prosthetics for amputees.

Many scientists, including physicist Stephen Hawking, expect that technology advancements will cure our most debilitating illnesses, but in doing so it may substantially change our human form. This rapid pace of developments compels one to consider whether we will eventually connect our brains directly to computers to constantly upload new knowledge and abilities directly into our minds. Perhaps, we may someday be directly interfaced with each other, like a giant World Wide Web of super-consciousness. Such possibilities ignite people's imaginations, as popular science fiction frequently depicts people merging with machines in many ways as cyborgs, superhumans, or genetically modified large-brained descendants of humans. Television has its super powered Bionic Man, while the "Borg" from *Star Trek* portrays the negative consequences of us merging with technology. In the stories of William Gibson, people directly interface their brains with a virtual web, which he terms the Matrix, inspiring the movies of the same name. As we redesign ourselves, we may become "transhuman," as futurist F. M. Esfandiary termed it in 1989. Many thinkers and scientists, including AI expert Marvin Minsky, nano-technology expert Eric Drexler, and the software pioneer and futurist Ray Kurzweil embraced this trend, with many calling themselves "transhumanists."

Transhumanists think of technology as part of our evolutionary continuum that will actually transform humanity in such fundamental ways that we would become an entirely new species. Pundits such as Ray Kurzweil proposed that, as computers become more capable with larger memory and as bio-scanning technology improves significantly in resolution, we will be able to digitize ourselves (at least our essential patterns), allowing us to live in virtual reality, and thus take whatever form we like. Hans Moravec and many others further contend that we will use genetics and neural interfaces to augment our intelligence in ways that will change the nature of our thought to the point where we might not even be able to relate to our descendants. Although these ideas themselves may seem like science fiction, many of the proponents and originators of these thoughts are scientists, actively planting the seeds of the technology in advanced research projects.

In this chapter, we will review the state-of-the-art of smart prosthetics, exoskeletons, and other assistive devices that are used for aiding, enhancing, and improving the physical capability and performance of human users and the disabled in particular. These advances are also greatly benefiting the technology of humanlike robots, since it is becoming easier to make their parts more lifelike.

SMART PROSTHETICS, EXOSKELETONS, AND ASSISTIVE DEVICES

Progress in developing smart prosthetics, exoskeletons, and assistive devices is continually being made in order to help disabled people regain their lost physical capabilities. These capabilities also benefit the development of humanlike robots. As needed, the developed devices are used to support, enhance, augment, or impede the physical movements of humans. Such devices are made as wearable, attachable, or even meant to be sat on. Also, there are now human–machine interfaces, where a disabled person is wired or implanted with artificial devices that allow restoration of the person's lost functions.

Development in interfacing humans to machines directly from their own brain, effective actuators and sensors, control algorithms, lightweight materials, and many other related technologies are resulting in unprecedented changes. Walking chairs are being developed to replace wheelchairs and carry humans in complex terrains such as up and down stairs and traversing areas that have various obstacles. These types of assistive devices are targeted for applications in medical rehabilitation, sports training, the heavy lifting industry, in-space activities, and military operations. A visionary concept for the use of an impeding exoskeleton is shown in Figure 4.1. This exoskeleton was proposed as an exercise machine for astronauts, to assist in reducing their bone and muscle deterioration that occurs in zero gravity during long missions. For health caregivers, Human Muscle Enhancers (HME), or Human Muscle Impeding (HMI) systems, can be used to reduce the cost of caring for the senior population, which is significantly on the rise worldwide. Moreover, such systems are needed for patients that are suffering from debilitating diseases or in a rehabilitation phase. These

Figure 4.1. A graphic illustration of an exoskeleton concept for zero-gravity countermeasures, using an on-demand motion impeding system as a wearable exercise machine. The image of the robot with the exoskeleton is a courtesy of Constantinos Mavroidis, Northeastern University, Boston, Massachusetts. The background image of the two astronauts is a courtesy of NASA.

patients may need ambulatory aids to sit, stand, or walk when it is not possible for them to use a cane, crutches, or walker.

The ability to augment motion is needed for other applications, too, including police operations or firefighter search and rescue missions. In certain emergency situations, it becomes necessary to run fast or lift heavy weights at capabilities that are beyond an average human. Military personnel, particularly those serving in combat areas, increasingly need to carry large loads and to traverse great distances under less than ideal conditions. The use of assistive devices can benefit industry greatly, too. Workers who lift heavy materials or equipment are at risk of lower back injuries, resulting in work time loss, workmen's compensation insurance premium increases to their employers, and reduced employee productivity.

ARTIFICIAL ORGANS AND NEURAL INTERFACING

In recent years, a significant body of research has been dedicated to the development of artificial organs and neural interfacing of these organs, and there have been some remarkable successes. Artificial heart valves, such as the St. Jude prosthesis, are now common in thousands of people. Kidney dialysis is a routine procedure in many hospitals, and artificial knees and hips have also become common. Complete artificial hearts, such as the Abiocor Implantable Replacement Heart, are being used to save lives while patients are waiting for a donor to provide a heart transplant.

The more complex or demanding the organ, the more difficult it is to make a prosthetic replacement. Specifically, the nervous system is the most difficult to replace with an artificial alternative. It is a very complex process to read and interpret signals directly from the nervous system and an even more challenging task to transmit signals into the nervous system in ways that we can interpret and utilize. For these reasons, the research into neuronal-prosthetics is particularly impressive, and it includes the development of retinal and cochlear implants, restoring lost senses, and brain stimulators that directly affect brain functions. Researchers, such as Okumura and Luo, reported the ability to use primate brain signals to control the operation of mechanical prostheses. This progress led to artificial arms and hands that are neurally interfaced, which for a long time was considered an impossible task.

Research seeking to interface the human brain and machines is underway in many academic and medical institutes, including Caltech, Duke University, MIT, and Brown University. Scientists at Duke University connected electrodes to the brain of a monkey and, using brain waves, the monkey operated a robotic arm, both locally and remotely via the Internet. Microelectronic chips were developed to recognize brain signals for movement and convert them into action. Monkeys fitted with such chips were trained to move a cursor on a computer monitor, where the chips translate signals from the brain's motor cortex, the region that directs physical movement.

Improvements in human–machine interfacing have reached the point in the United States that the Food and Drug Administration (FDA) was able to approve, in 2005, on a limited basis, the conducting of such experiments on humans. To this end, Cyberkinetics, in Foxborough, Massachusetts, implanted microchips in the motor cortex region of the brains of five quadriplegic patients that allowed them to control the mouse of a personal computer. The near-term objective of this study is to develop neural-controlled prosthetics. The recently developed chips last up to a year, and efforts are being made to develop a longer-lasting wireless system.

The ability to control prosthetics requires feedback in order to provide the human user with a "feeling" of the environment around their artificial parts and the loads on their various artificial components. Tactile sensors, haptic devices, and other interfaces can be used to deliver such feedback. The sensors are needed to help users protect their prosthetics from potential damage (heat, pressure, impact, etc.), just as we have the capability to protect our biological limbs. Kuiken and his co-investigators from North-western University in Chicago have shown that transplanting the nerves from an amputated hand to the chest allows patients to feel sensation in their artificial hand. This capability may lead to the development of prosthetic arms with sensors on the fingers that will transfer tactile information from the device to the chest, possibly making a user feel sensation in a prosthetic hand similar to a naturally felt sensation.

Vision and hearing are critical sensors for humans and machines. They are used for communications, determining the conditions of the surrounding environment, avoiding obstacles, identifying dangers, and obtaining other important information. The ability to Interface the human brain with visualization and hearing devices has already been successfully demonstrated. Increasingly, hearing devices are implanted in patients. However, imaging devices are still at the research stages.

RESTORING VISION

Generally, the capability of the human eye, its iris, and the eyelid is being emulated in today's cameras. However, in spite of the progress in camera technology, the eyes and the human brain have far superior capabilities, including fast and reliable image interpretation and recognition, the ability to rapidly focus without moving the lens location in the eye, extremely high sensitivity, and the ability to operate in a wide range of light intensities, from very dark to quite bright light as well as rapidly responding to changes in intensity. However, artificial vision capability has grown significantly with the miniaturization and increased resolution of digital cameras. Such cameras are now part of webcams for telecommunication via computers and are used in most cellular phones. For vision-impaired humans, researchers are working on creating implants that can help such persons regain their ability to see. To make robot and other vision-aided systems effective, the developed cameras need to provide three-dimensional images in real-time, with performance that is as close to the human eye as possible.

Sophisticated visualization and image recognition capabilities are increasingly being developed for use in security systems. However, although lab demonstrations have been very successful, these systems still have an insufficient rate of recognition success. Once the reliability issues are overcome, this technology will likely become a standard support tool as part of homeland security systems in airports, public areas, or even in our homes.

Figure 4.2. An illustration of an intraocular retinal prosthesis showing a glasses-like imaging device (on the left) and a human eye with an implanted artificial retina (on the right). This image was created by Adi Marom and was inspired by the system that was developed at the University of Southern California.

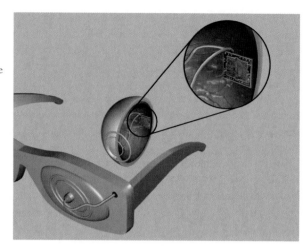

In the treatment of vision deterioration, one success to date involves a retinal implant in conjunction with an external miniature video camera mounted in a pair of spectacles and a wireless receiver. The prosthesis (see Figure 4.2) uses an array of electrodes, implanted directly into the retinal nerves, to transmit signals from a video camera to the brain. Patients who were equipped with such a system have reported the ability to distinguish between light and dark as well as between objects and detect motion. Although researchers are exploring the use of a silicon retina that would function without an external camera and receiver, the primary difficulty lies in developing materials that would be compatible with the composition of the eye itself. In a different approach, the retinal implant is replaced by a brain implant that uses an external camera and electrodes attached to the visual cortex.

RESTORING HEARING

Unlike retinal implants, which are in their infancy and not approved yet for general use, the cochlear implants are a relatively mature neural interface technology, first developed in 1957 and today used by over 100,000 people. As opposed to hearing aids, cochlear implants repair hearing abilities by directly stimulating nerves within the cochlea, thus restoring hearing in profoundly deaf people. The cochlear implant takes audio signals from a microphone and translates them into electrical signals that stimulate the nerves. Usually a patient must acclimate to the device that is connected to the nerves and over time learns to interpret the alien signal as sound. Early cochlear implants transmitted a narrow segment of sound, usually wavelengths of sound associated with speech, thereby restoring auditory verbal communication. However, generally, they did not restore the complete hearing capabilities. Today, these devices are providing impressive hearing restoration.

SMART PROSTHETICS

Prosthetics shaped like the human hand date back to ancient Egypt. Artificial hands with mechanical movement capability were already in use during the early part of the twentieth century (see an example in Figure 4.3). These hands were designed to allow mechanical control of the artificial fingers by moving the healthy part of a severed arm. The hands were made of materials that were intended to match the appearance of a human's real hand, but given the limited technology of many decades ago, the hand was quite rigid, and the color did not match well. Advances in recent years allowed significant improvements in prosthetics, making them more flexible and easier to control as well as appear very much like natural human hands. An example of such a hand (made by Hanger Inc.) is shown in Figure 4.4, where Carrie Davis, who is a National Upper Extremity Patient Advocate for Hanger Prosthetics and Orthotics, is

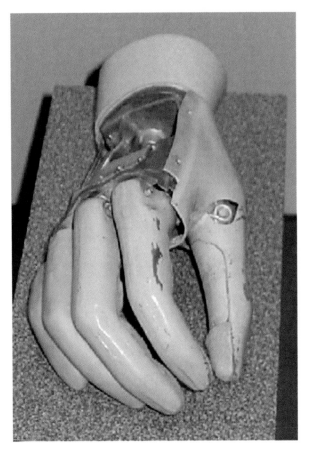

Figure 4.3. A mechanical hand for use as a prosthetic. Photo by Yoseph Bar-Cohen at the Smithsonian Museum in Washington, DC.

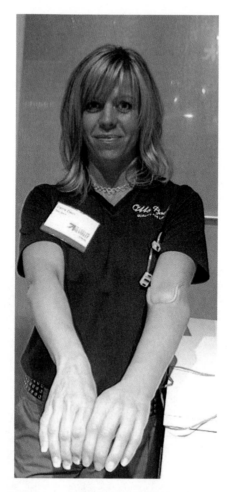

Figure 4.4. Carrie Davis is showing her natural right hand and prosthetic left hand. Photo by Yoseph Bar-Cohen at the 2007 NextFest Exhibition of *Wired* magazine.

showing her natural right hand and prosthetic left hand, demonstrating the success of the latest technology at the Wired Magazine's 2007 NextFest exhibition in Los Angeles, California.

In the late twentieth century, scientists began incorporating advanced biology research and robotics technology into prosthetic devices, in order to emulate the performance of actual limbs and enable better control of prosthetic devices. The results included myoelectric arms, which helped the artificial limbs of amputees to grasp objects, though with limited control (Figure 4.5). The latest research in robotic prosthetics has resulted in much more complex arms that are controlled directly by the brain of the human user. Such arms were developed under a DARPA program in the last 4 years and are described later in this chapter.

Figure 4.5. A modern myoelectric hand used to perform physical tasks. Photo courtesy of Otto Bock Healthcare.

One of the leading proponents of the research related to smart prosthetics is the U.S. military research funding organization known as DARPA. The research was motivated by the number of hand and arm amputations sustained by soldiers in Iraq and Afghanistan. Many of these soldiers preferred the body-operated "hook" replacement, which has been in use since World War I, over the use of prosthetic devices that are controlled via electrodes placed on the skin to read muscle signals. The focus of the DARPA's research is primarily on the development of a neurally controlled artificial limb to restore full motor sensory capability to upper extremity amputee patients. The program is called Revolutionary Prosthesis, and its objective is to lead to the development of prosthetics that are controlled by the human user and perform, feel, and look like a person's original limb. Moreover, the developed prosthesis needs to have the sensitivity and strength to handle the minute tasks of daily living.

To meet the DARPA objectives, techniques are developed to directly wire artificial limbs into the nervous system, rather than relying on the muscle-controlled technology (myoelectric) that is widely used. To provide the user the required control it is necessary to sense the electrical activity that it is intended to activate the related muscles in order to operate the prosthesis. Generally, the control techniques are limited in capability and do not provide the sensation of what the prosthesis is touching. The complexity of sensing nerve signals makes it a daunting task to directly communicate to and from the nervous system in a way that is intuitive to the user. These signals are generally called "efferent" signals, as they radiate out from the nervous system, and they can be quite subtle and difficult to detect and use. Communicating sensations back to the nervous system ("afferent" axons) is also a complex task, requiring interface of the electronic signals into the nerves in a way that nerves can interpret. The task of developing prosthetics that have the power and motor complexity of human limbs in a package that is mobile and lightweight is extremely challenging.

DARPA started with an intermediate goal of "increasing the range of motion, strength, endurance, and dexterity of upper extremity prosthetic devices." To support this effort, several prototypes were developed, and one of them is shown in Figure 4.6. Also, success has been achieved in rerouting nerves from an amputated limb and muscles in an area of the body that was not affected by the amputation. The most successful work took place at the Neural Engineering Center for Artificial Limbs (NECAL) in the Rehabilitation Institute of Chicago (RIC). At this center, rerouting sensory nerves provided a measure of tactile feedback through the prosthesis.

The concept of cognitive prosthetics may seem to be a very futuristic idea. However, early research is already underway into neural prostheses for the treatment of various medical conditions, including the treatment of Parkinson's disease and the control of epileptic seizures. The neural interfacing capabilities to date are allowing transfer of relatively low bandwidth information into and out of the nervous systems of people. More complete interfacing of our technology to transfer large quantities of data, such as literature, images, or ideas, will require further improvements in interfacing capability and greater understanding of the function of the human brain. Yet, the recent pace of advancements encourages one to dream of a time in our lives when we solve many of these problems and become able to upload new memories and skills as easily as downloading software onto a PC from the Internet.

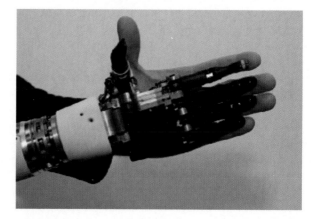

Figure 4.6. Robot hand developed under a DARPA program. Photo courtesy of Dirk Van Der Merwe, Deka Research & Development Corp., New Hampshire.

PROSTHETICS THAT GIVE HUMANS SUPERIOR ABILITIES

Progress in making smart prosthetics led to capability levels that allow disabled persons to use their artificial limbs to perform highly demanding functions, such as running, climbing, and so on. An example of the use of passive prosthetics in a running race is shown in Figure 4.7, where a disabled athlete (photo by Fatronik, San Sebastian) with one prosthetic leg can be seen preparing for a race. As improvements in materials, actuators, microprocessors, and miniature batteries continue to increase, it is becoming easier to produce smart and highly effective prosthetics that look very much like natural organs.

The improvements in developing prosthetics have been enormous, and some of the disabled are now able to walk or run as good as or even better than humans with natural legs. An example that highlights the progress in the development of prosthetics is Oscar Pistorius, a South African with two prosthetic legs who sought to participate as a runner in the 2008 Olympics games in China. Initially, his request was turned down, because he was considered to have a competitive advantage. In rejecting his request to run, the International Association of Athletics Federations (IAAF) made the following statement: "An athlete using this prosthetic blade has a demonstrable mechanical advantage

Figure 4.7. A disabled athlete equipped with a smart prosthetic leg preparing for a running race. Photo courtesy of Fatronik, San Sebastian.

when compared to someone not using the blade." This decision was reversed on May 16, 2008, after an appeal that was submitted to the Court of Arbitration for Sport (CAS), which is the sport's highest tribunal. In overturning the ban imposed by IAAF, CAS unanimously ruled that this 21-year-old South African was eligible to race against able-bodied athletes. In deciding to overturn the ban it was stated that (based on tests conducted by a team led by MIT) IAAF failed to prove that running blades give an advantage over the natural legs of able-bodied athletes. Unfortunately, Oscar Pistorius ended up not participating in the 2008 Olympics in China. In mid-July 2008 the track and field body of South Africa confirmed that he had not made the requirements for their 400 m relay team. Although Oscar Pistorius did not participate in the Olympic competition, the fact that he was given the opportunity to compete against runners with natural legs suggests a major milestone in the performance of prosthetics today.

In an effort to take advantage of the progress made in the area of prosthetics and support disabled soldiers, DARPA initiated the program called "Revolutionizing Prosthetics." Specifically, the objective of this program is to develop a smart prosthetic hand that can operate and feel as good as real limbs. Thanks to this program we now have smart prosthetic hands and arms that match quite well the appearance and functions of a real human hand. Photos of such hands are shown in Figures 4.8 and 4.9. The hand that is shown in Figure 4.9 is activated by movement of muscles in the healthy section of this soldier's right arm. The capability of a full arm is shown in Figure 4.10, where a prototype was used to support reading a book just the way a human hand can do. Another arm that was developed under the DARPA's Revolutionizing Prosthetics program is the one shown in Figure 4.11; it is mounted on a user and is operating a drill. The prosthetic hand of this arm is controlled via a joystick worn on the healthy section of the severed arm. Since there are a limited number of movements that can be

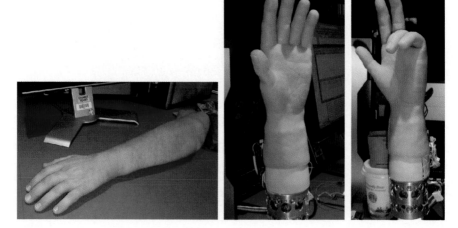

Figure 4.8. Lifelike prosthetic hands and arm. Photo by Yoseph Bar-Cohen at the 2007 DARPATech that was held in Anaheim, California, in August.

Figure 4.9. Smart prosthetic hand demonstrated by Jonathan Kuniholm (a veteran who was injured in Iraq) at the 2007 DARPATech. Photo by Yoseph Bar-Cohen at the 2007 DARPATech.

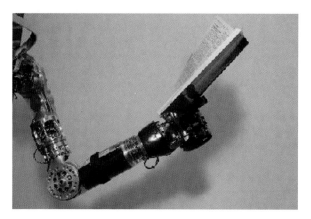

Figure 4.10. A humanlike arm holding a book in a reading pose, illustrating the ability to operate as an effective prosthesis. This arm was developed under the DARPA program Revolutionizing Prosthetics. Photo courtesy of Dirk Van Der Merwe, Deka Research & Development Corp., New Hampshire.

produced by the user's hand, the raising and lowering of the prosthetic arm is done by mirroring the movements of the user's tow using a related set of sensors. The control via the hand and tow requires the user to practice the harmonious operation of the prosthetic arm, and with sufficient hours of use the task becomes quite smooth and the

Figure 4.11. A prosthetic arm that is controlled by mirroring movements of the wearer's severed arm and the tow. This arm was developed under the DARPA. Photo courtesy of Dirk Van Der Merwe, Deka Research & Development Corp. New Hampshire.

movements appear natural. One may want to remember that the use of our hands and legs took us many months to learn as infants, so training over tens of hours is a small price to pay for such sophisticated capabilities using such a simple interface.

An alternative to the "joystick"-type operation of prosthetics is to develop a natural control directly from the user's brain and nervous system. At the RIC, using the capabilities that were developed under the DARPA, several disabled persons were "rewired." The nerves of their severed arm were connected in the chest muscles to a prosthetic arm. The nerves were the ones intended to operate the missing hands, and these persons were able to move their mechanical arm by creating contraction via the actuators that are stimulated by the related nerves of the specific muscles. In Figure 4.12, the world's first person that was wired to activate such a wired prosthetic hand is shown.

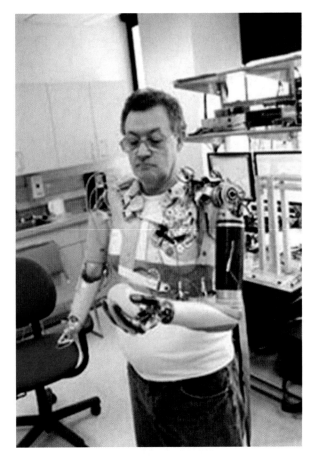

Figure 4.12. Jesse Sullivan is the world's first person to have a robotic prosthetic arm driven by his own nerve impulses from his brain. Photo courtesy of the Rehabilitation Institute of Chicago.

His name is Jesse Sullivan (from Dayton, Tennessee); when this experiment took place, Sullivan was 50 years old. He lost both his arms in a power-line accident. A close-up of one of the developed prosthetic hands is shown in Figure 4.13 performing a complex task of inserting various shapes through the matching holes in a board with a matrix of various configurations.

Similar to the objectives of the DARPA initiative, a group of scientists and engineers at the European Union are working on a hand called Cyberhand. This group is seeking to develop a highly dexterous artificial hand with a sensory system to provide patients with active feeling. The development of the Cyberhand started in 2002, with the intent to increase knowledge about neural regeneration and sensory-motor control of the human hand and to develop a realistic and effective prosthetic hand. A photo of the developed Cyberhand is shown in Figure 4.14.

Figure 4.13. One of the prosthetic hands (left) developed under DARPA, being used to perform a complex task. Photo courtesy of the Rehabilitation Institute of Chicago.

Figure 4.14. The Cyberhand, a European artificial hand. Photo courtesy of Paolo Dario, Director, Polo Sant'Anna Valdera, Scuola Superiore Sant'Anna in Pisa, Italy.

MORE ABOUT EXOSKELETONS

Another type of device that is considered part of the group of robotic-related mechanisms is the exoskeleton. As an example, the image in Figure 4.15 shows a schematic illustration of an exoskeleton mounted on a graphic simulation of a user. This exoskeleton was proposed to enhance or impede leg movements. Arrows show the direction of actuation for the right leg. As human muscle enhancing systems, exoskeletons are used to aid human users

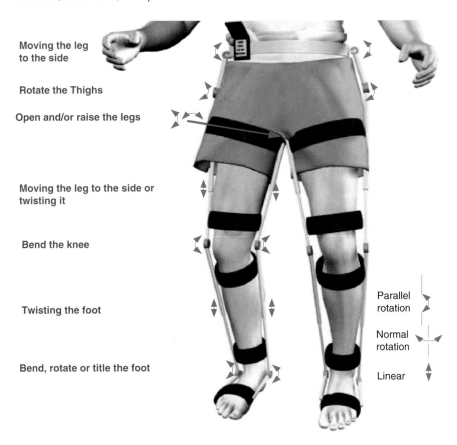

Moving the leg
to the side

Rotate the Thighs

Open and/or raise the legs

Moving the leg to the side or
twisting it

Bend the knee

Twisting the foot

Bend, rotate or title the foot

Parallel
rotation

Normal
rotation

Linear

Figure 4.15. A schematic view of an exoskeleton mounted on a graphic simulation of a user.
Arrows show the direction of action for the right leg. This conceptual exoskeleton was developed
by Yoseph Bar-Cohen and Constantinos Mavroidis. Photo courtesy of Constantinos Mavroidis,
Northeastern University, Boston.

perform improved physical tasks with a capability that is far superior to their own.
Generally, exoskeletons are developed for such applications as medical rehabilitation, sports
training, and combat operations of military personnel. Exoskeletons can reduce cost by
providing an alternate option to a health caregiver in situations where an elderly person has
control and can walk a little. The use of human muscle enhancing systems can also
minimize the number of human helpers that are needed by providing partial support to
the patient. Also, such systems can benefit patients with debilitating muscular weakness,
including senior citizens, patients confined to a lengthy bed rest, and patients with spinal
nerve damage. Military personnel in the field can benefit greatly if they have a greater load-
carrying capability with less strain and the ability to ambulate further distances than present
conditions allow. Industry can also gain benefits. Industrial workers who lift heavy materials
or equipment experience many lower back injuries, resulting in work time loss, diminished
employee productivity, a financial burden on the health care industry.

Several models of exoskeletons have been developed in labs worldwide. An example of such an exoskeleton includes the one called Berkeley Lower Extremity Exoskeleton (BLEEX), which was developed at the University of California, Berkeley. This machine consists of rigid motorized mechanical metal leg braces that are attached to the user's feet, and it includes a power unit that is carried as a backpack-like frame. The development of the BLEEX was funded by DARPA with the objective of enhancing the capability of army medical personnel to carry injured soldiers off a battlefield; fire-fighters to haul their gear while climbing many stairs to put out fire in high-rise buildings; or rescue workers bring food and first-aid supplies to areas where vehicles cannot enter. Another example includes the exoskeleton developed by Yoshiyuki Sankai and his colleagues at Tsukuba University, Japan, with the objective of helping

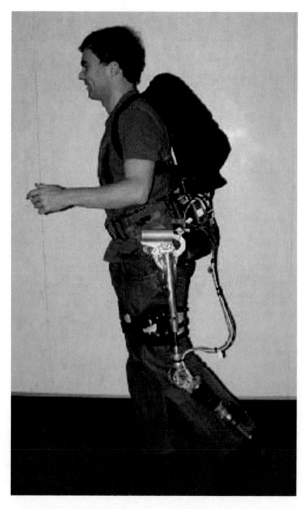

Figure 4.16. A motorless exoskeleton using variable damping in a passive system. Photo courtesy of Hugh Herr, Media Lab, MIT.

people struggling to walk. Their Hybrid Assistive Limbs, also known as "bionic trousers," senses motor nerve signals and moves in coordination with the user's body to augment his or her movement. This suit is currently being tested for use with stroke victims and patients with spinal cord damage.

To simplify the operation of exoskeletons, researchers at MIT developed a motorless system that uses minimal power and can carry about a 36 kg (80 lb) load (see Figure 4.16). This exoskeleton was developed under the lead of Hugh Herr, with the objective being to design mechanical structures that transfer much of the load directly to the ground rather than via the walker's leg. This passive exoskeleton system is quieter than the powered types. It has elastic energy-storage devices at the ankle and a variable damping device at the knee joint. Springs are used to propel the legs forward on the next stride, while a damper lets the leg swing freely as it moves forward. Once the heel strikes the ground the damping increases to prevent the knee from buckling under the weight of the payload. The limited power that is used drives the variable damping mechanism, and when walking it can reach a power consumption of about 2 W. One of the deficiencies of this exoskeleton is its limitation in operating in complex terrains such as on stairs and on uneven surfaces.

LEGGED CHAIRS

Although exoskeletons are developed as wearable structures that augment the abilities of humans or help with rehabilitation efforts, there are also now walking chairs, where the user sits on the device, which consists of a chair with two legs. These legged chairs are intended to serve as a superior alternative to wheelchairs, since the latter require flat surfaces for mobility and are difficult or even impossible to operate in complex terrain. As mentioned in Chapter 2, Toyota is developing such a chair as part of their Mountable Partner Robots program. The

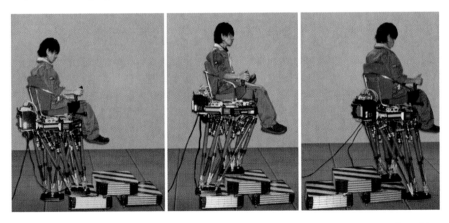

Figure 4.17. The walking chair shown above is named the Human-Carrying Biped Walking Robot (developed at Waseda University, Japan) and is capable of carrying humans over difficult terrain and climbing stairs. Photo by Yoseph Bar-Cohen at the 2007 NextFest Exhibition of *Wired* Magazine. The background of the photo was replaced by graphic artist Adi Marom to show a wall and a floor, making it easier to see what was demonstrated.

device that they developed is capable of carrying its user anywhere he or she needs to go. Another group that is developing such an ambulation chair is the Biped Locomotion Department of Waseda University in Tokyo, Japan. Their legged chair is called Human-Carrying Biped Walking Robot, and it is capable of carrying humans that weigh up to 77 kg (170 lbs) using a chair that weighs about 61 kg (135 lbs). This chair-robot has a pair of 122 cm (4 ft) tall legs that can move on uneven terrain and can climb stairs. A demonstration of climbing up and down over a two-step structure was demonstrated recently at the *Wired* Magazine 2007 NextFest Exhibition that was held in Los Angeles, California (see Figure 4.17).

FUTURE DEVELOPMENTS

Cyber Love

As discussed earlier in this chapter, prostheses are starting to provide abilities that are nearing those of an organic limb. As prosthetics are developed that surpass the capability of our own limbs, new issues and questions will be raised about the future direction of the technology. With significant number of body parts made of artificial organs one may wonder how we would relate to each other even in such areas as romance. Recent advances in neuroscience have unlocked in greater detail the neurochemistry of love. It has been shown that manipulating the neurotransmitters dopamine and oxytocin inspires feelings of love and romance in animals and people. Will there be a day where we take pills to reinforce our deep feelings of love, to help us grow closer in marriage and other relationships? It remains to be seen whether the effects of such chemical manipulations would be as deep and meaningful as those that emerge spontaneously between people. In the meantime, consider that another neurotransmitter of love, phenylethylamine (PEA), is an ingredient in chocolate, which may be activated merely by eating the sweet treat.

One may also consider the effects of love upon our machines. As machines become more humanlike, and we are able to interface with machines in an ever more integral manner, how will we begin to feel towards our machines? Will we fall in love with our robots and computers? Will they fall in love with us? As we merge with our technology, will there be a line that distinguishes us from our technology? Not only can we anticipate that such mergers may change our sense of relationships, but also the concept of high bandwidth communication between human minds promises, in effect, to fuse one's mind with that of a loved one. The issues that are associated with the use of robots in physical interactions with humans, including sex and others, are not covered in this book, but can be found elsewhere in the literature.

Cyborgs

In 1960, Manfred Clynes and Nathan Kline coined the term "cyborg" to describe an organism that functions and consists of an independent combination of artificial ("cybernetic") and natural ("organic") parts. Initially, this term was used in conjunction with the idea that humans would need technological enhancements for operating in the environment of space. However, the term has evolved to include even a person using a common device such as a pacemaker. As per this definition, a cyborg is usually a human being with machine or machine-controlled parts. But the definition can also be expanded to include collective entities such as corporations or cities that have individual elements sharing communications and control. Cyborgs can be divided into two types, the restored and the enhanced. The restoration of a cyborg returns a damaged individual to a normal state. The enhancement of a cyborg creates an individual who functions at a level superior to the normal state.

The trends in robotics, biotechnology, and nanotechnology call our attention to issues of enhancement in particular, issues related to superhumans. The functionality of technologically enhanced people may so radically outpace the capability of their unenhanced peers that eventually they will be seen practically as a separate species. A host of ethical questions are expected to arise when considering the use of technology just for enhancement, rather than merely to repair illness or injury. The use of technology in this way has been dubbed "enhancement-prosthetics."

Biologist Julian Huxley first used the term "transhumanism" in 1957 to mean "man remaining man, but transcending himself, by realizing new possibilities of and for his human nature." In 1990, the science philosopher Max More defined transhumanism as "a class of philosophies that seek to guide us towards a posthuman condition." He continues: "Transhumanism shares many elements of humanism, including a respect for reason and science, a commitment to progress, and valuing of human (or transhuman) existence in this life. Transhumanism differs from humanism in recognizing and anticipating the radical alterations in the nature and possibilities of our lives resulting from various sciences and technologies". The term transhumanism received its formal definition in 1999 by the World Transhumanist Association as:

1. The intellectual and cultural movement that affirms the possibility and desirability to fundamentally improve human conditions through applied reason, especially by developing and making widely available technologies to eliminate aging and to greatly enhance human intellectual, physical, and psychological capacities.
2. The study of the ramifications, promises, and potential dangers of technologies that will enable us to overcome fundamental human limitations, and the related study of the ethical matters involved in developing and using such technologies.

Transhumanists believe that human enhancement through external devices represents a natural step forward in the progression of evolution. In this way, the transhumanists argue that it is imperative that we use technology to assist the human race in growing beyond its natural limitations, limitations of strength, speed, and intelligence. This may also include counteracting the biology of aging by forestalling or reversing its effects. The use of technology for enhancement is controversial, as many believe that it's inherently wrong to become "unnatural" and that we may be opening a Pandora's box of problems that will ultimately harm humanity. Others may object to the use of such technology based on religious principles. One may also wonder about the potential consequences of such developments if we tinker with our nervous system without understanding completely the potential consequences.

Rushing into the technology is tempting, since it may boost our abilities to obtain competitive advantages such as we find in the lucrative arena of competitive sports.

Even as the use of steroids in sports has come under increasing attack, athletes continue to use performance-enhancing technologies as soon as they become available. Many gray areas arise as athletes apply technologies. Already, some athletes are using designer proteins, smart drugs, and gene therapy to get the edge to reach the top of their field. In science fiction, the consequences of such efforts are almost always disastrous, as when Bruce Banner is transformed into the Incredible Hulk in the TV series and movies that carried this name.

Although many transhumanists believe that technological and scientific advances should be used to address social ills, transhumanism focuses on the enhancement and improvement of individual human bodies by the use of technology. In transhumanistic thought, the idea of self-evolution replaces that of natural evolution. By means of ever-emerging technologies, the individual creates his or her own evolution that surpasses the evolution natural to the species. Technological descendants of humans may not be humans at all, even in part, and so might be termed instead "transbemans." The term is derived from "being" as an existential essential of consciousness, affording any conscious being the rights we would traditionally reserve for humans.

Transhumanism has critics, too, who claim that it is not likely that its goals can ever be achieved and that there are many moral and ethical implications to its implementation. This criticism of transhumanism suggests that its claims are based on flawed assumptions about future advances in technologies and human adaptation to them. For example, we may find ourselves on a slippery slope – dependent on technologies that transform us. It may be useful for us to consider now how we would craft the technologies to enhance our positive traits and to avoid the potential nightmarish or dehumanizing effects of any technology trends.

SUMMARY

The need to help the disabled by developing smart prosthetics that have the appearance of and perform the functions of missing limbs is driving the development of increasingly better devices. Academic institutes, industry, the military, and many other organizations and groups are involved with the development of prostheses that can be controlled directly by the user's brain, and success is already being implemented in artificial appendages. Further, exoskeletons and legged ambulation mechanisms are being developed to enhance a person's ability to perform tasks, to carry heavy weights, and traverse great distances rapidly with minimal effort. Also, there are chairs that are able to perform far better than wheelchairs in carrying people and ambulating in complex terrains, including on stairs and on uneven surfaces.

One can speculate endlessly about the effects that today's technology trends may have on humanity. However one thing is certain: technology is increasingly modifying and enhancing the capability, appearance, and behavior of people, and the progress being made is not going to slow down anytime soon. Over the last century, our technology has emerged from the science fiction of speculation to actual smart prosthetics and functioning bionic robotic enhancements of people. These technologies have added greatly to our capabilities, and each day brings new advances. Some may argue that it is preposterous to presume that humanity will be transformed, but we need to be aware of the potential consequences to guide the future development of our enhancement technologies.

BIBLIOGRAPHY

Books and Articles

Agrawal, R. N., M.S. Humayun and J. Weiland, "Interfacing microelectronics and the human visual system," Chapter 17 in Bar-Cohen, Y., (Ed.), *Biomimetics–Biologically Inspired Technologies*, CRC Press, Boca Raton, FL, (Nov. 2005), pp. 427–448.

Bar-Cohen, Y., C. Mavroidis, M. Bouzit, C. Pfeiffer, and B. Dolgin, "Remote MEchanical MIrroring using Controlled Stiffness and Actuators (MEMICA)," NTR, January 18, 1999, Item No. 0237b, Docket 20642, January 27, 1999. Rutgers U. Docket #99-0056, Filed patent on (September 11, 2000).

Clynes, M. E, Kline, N. "Cyborgs and Space," *Astronautics Journal,* American Rocket Society Inc, New York, N.Y., (September, 1960), p. 33.

Drexler, E. *Engines of Creation: The Coming Era of Nanotechnology.* Anchor Books (1986), p. 234.

Esfandiary, F. M. (named himself Fm-2030), *Are You a Transhuman?: Monitoring and Stimulating Your Personal Rate of Growth in a Rapidly Changing World,* Warner Books, Clayton, Australia, (January 1989)

Fisch, A., C. Mavroidis, Y. Bar-Cohen, and J. Melli-Huber, "Haptic and Telepresence Robotics" Chapter 4, in Y. Bar-Cohen and C. Breazeal (Editors), *Biologically-Inspired Intelligent Robots*, SPIE Press, Bellingham WA, Vol. PM122, (2003), pp. 73–101.

Gibson, W., *Neuromancer,* Ace Books, New York, (1984).

Graham-Rowe, D., "MIT Exoskeleton Bears the Load," *Technology, MIT Review,* (September 26, 2007).

Gray, C. (Ed.), *Cyborg Worlds: The Military Information Society,* Free Associations, London, (1989).

Haraway, D., *Simians, Cyborgs, and Women: The Reinvention of Nature,* Free Associations, London, (1991).

Harris, J., *Wonderwoman and Superman: The Ethics of Human Biotechnology,* Oxford University Press, Oxford, (1992).

Hughes, J., *Citizen Cyborg: Why Democratic Societies Must Respond to the Redesigned Human of the Future*, Westview Press, (2004).

Huxley, J. "Transhumanism" in *New Bottles for New Wine*, London: Chatto & Windus, (1957), pp. 13–17.

Kazerooni, H. and Guo J., "Human Extenders", *Transactions of the ASME, Journal of Dynamic Systems, Measurements, and Control*, Vol. 115, (1993), pp. 281–290.

Klugman, C. M., "From Cyborg Fiction to Medical Reality," *Literature and Medicine*, Vol. 20, No 1, (Spring, 2001), pp. 39–54.

Kuiken, T. A., P. D. Marasco, B. A. Lock, R. N. Harden, and J. P. A. Dewald, "Redirection of cutaneous sensation from the hand to the chest skin of human amputees with targeted reinnervation," *Proceedings of the National Academy of Science*, USA, 10.1073/ pnas.0706525104, (November 28, 2007).

Kurzweil, R., *The Age of the Spiritual Machines*, Viking Press, New York (1999).

Kurzweil, R., *The Singularity Is Near: When Humans Transcend Biology*, Viking Press, New York, (2005), pp. 180, 450.

Levy, D., *Love and Sex with Robots*, HarperCollins Publishers, New York, (November 6, 2007)

Mavroidis, C., Y. Bar-Cohen, and M. Bouzit, "Haptic Interfacing via ElectroRheological Fluids," Topic 7, Chapter 19, *Electroactive Polymer (EAP) Actuators as Artificial Muscles – Reality, Potential and Challenges*, Y. Bar-Cohen (Ed.), 2nd edition, Vol. PM136, SPIE Press, Bellingham WA, (2004), Pages 659–685

McNamee, M. J., and S. D. Edwards, "Transhumanism, medical technology and slippery slopes," *Journal of Medical Ethics*, Vol. 32, (2006), pp. 513–518.

Moravec, H., *Mind Children: The Future of Robot and Human Intelligence,* Cambridge, MA, Harvard University Press, (1988).

More, M., *Principles of Extropy*, http://extropy.org/principles.htm Accessed on 10-14-2007, (1990–2003).

Okumura, M., and Z. Luo. On NIRS-based Brain-Robot Interface, *IEEE Int. Conf. on robotics and biomimetics (ROBIO)*, (2007).

Pecson, M., K. Ito, Z.W. Luo, A. Kato, T. Aoyama and M. Ito, "Compliance Control of an Ultrasonic Motor Powered Prosthetic Forearm," *Proceedings of IEEE International Workshop on Robots and Human Communication* (Ro-Man'93), pp. 90–95, (1993).

Rothblatt, M., e-mail exchange of Y, Bar-Cohen and this philosopher (December, 2007)

Walsh, C., K. Endo, H. Herr, "A Quasi-Passive Leg Exoskeleton for Load-Carrying Augmentation," *International Journal of Humanoid Robotics*, Vol. 4, No. 3 (September, 2007).

Walsh, C., K., Pasch, H. Herr, "An autonomous, underactuated exoskeleton for load-carrying augmentation," *Proceedings of the IEEE/RSJ International Conference on Intelligent Robots and Systems (IROS)*, Beijing, China, (October 9–16, 2006).

World Transhumanist Association, *The Transhumanist Declaration,* website address: http://www.transhumanism.org/index.php/WTA/faq21/79/ accessed on 10-14-2007 (1999, 2002).

Internet Websites

Development of passive knee joint (in Japanese):
    http://www.assistech.hwc.or.jp/ASSISTECH/ru4/Stabilized_P.htm
DARPA's "Revolutionizing Prosthetics"
    http://www.darpa.mil/dso/solicitations/prosthesisPIP.htm
    http://www.jhuapl.edu/newscenter/pressreleases/2007/070426.asp
    http://www.msnbc.msn.com/id/13325828/site/newsweek/?GT1=8211

Chapter 5
Mirroring Humans

Humanlike robots are being developed to mirror us in many ways, including our artistic form, movement, and intelligence. The latter is still on a limited level, but it is progressing quite rapidly as improvements in artificial intelligence (AI) technology continue to be made. An illustration of the physical mirroring of a human is shown in Figure 5.1, where the natural human is mirrored with a humanlike robot. Mirroring humans involves not only a reflection of the physical image but also evokes intelligent mental models that help us understand the world around us. In his book *Godel Escher Bach,* AI researcher Douglas Hofstadter refers to such mirroring as "strange loops." The feedback between our intelligent technology, our memes, and the scale of our aggregate intelligence is helping to shape robots that exhibit increasingly lifelike behavior.

Overall, the science of robotics, which refers to the engineering, technology, and marketing of robots, is expanding at an accelerating rate. Software (including AI) is evolving rapidly, the cost of computers continues to drop, and alternatives to silicon transistors are promising to extend the changing trends into new media. These technologies can augment the intelligence of the human species, both in speed and quality. If we attempt to extrapolate these trends a few decades into the future, we may anticipate more than just machines that perfectly imitate human appearance and cognition. We may also expect organism-like machines that extend biological evolution into the nonbiological domains of silicon, nano-tubes, and other materials, machines that in some respects may grow superior to us cognitively, and possibly the fusion of ourselves with our machines. We may expect to experience direct neural interfacing with our computers, and to be network interfaced with other individuals into super groups that operate as single identities. Human individuals may be scanned, digitized,

Y. Bar-Cohen, D. Hanson, *The Coming Robot Revolution*, DOI 10.1007/978-0-387-85349-9_5,
© Springer Science+Business Media, LLC 2009

Figure 5.1. In many ways, humanlike robots will increasingly mirror natural humans.

and exist in the Web alone or rapidly evolve into lifelike robots. What would we become and what the world will turn into will depend on key actions taken before the change accelerates beyond our control.

Even though we cannot really predict the full range and specific changes that human-inspired robots may bring about, the power of our imagination may help to examine the coming impact of robots. The literature of science fiction can help us speculate about the future and theorize about the potential consequences. Certain stories stand out as guideposts in the evolution of robots. Shelley's *Frankenstein* (1818) transmits much of the cautionary aspects of ancient tales of the Golem, while presaging possible *Terminator*-like nightmares wherein the creation sets out to destroy systematically the world of their human creators. Asimov's robot stories (1942 and 1968) mirror the hopeful aspects of the old Golem tales, but his robots are more intelligent and inherently friendlier than previous automata depictions. Philip Dick's robot stories highlight the importance of compassion in defining humanity, be it artificial or biological humanity.

We must not forget, however, that the outputs of literature, art, technology, and philosophy are continuously progressing in what Dawkins refers to as the evolution of memes (sort of very rapidly evolving human software). When viewing science fiction as part of our memes, events in science fiction literature are helping us to reconsider what it means to be a human and possibly reinvent ourselves as humans. In a converse sense, humanlike robots themselves qualify as a form of fiction, being artificial portrayals of human characters. In fact, the personality of current robots can be considered a form of nonlinear theatrical narrative, science fiction incarnate. Unlike literary fiction dealing with robots, however, the physical robot is a fiction that can start to reinvent itself, bootstrapping rapidly toward a reality. Current robots may merely mimic humanlike intelligence in a ghostly manner – while nowhere near human level, they are undeniably smarter than any previous technology: making eye contact and conducting verbal conversation. We may possibly find in a mere 20 years or so that the fiction of robots has actually transformed into the true evolutionary progeny of the human species.

SOCIABLE ROBOTS

Emotional and sociable robots benefit greatly from having AI. Specifically, affective computing uses the natural emotions and expressions of humans to make computers more accessible to people, and make them think and reason more like people. In the subset of affective computing called social (or sociable) robotics, researchers are working to create human–computer interfacing devices that are capable of displaying the emotional inner state of a machine by mimicking nonverbal expressions native to human communications. The face serves as humanity's primary tool for expressing affective states, and the human nervous system is finely attuned to understanding the visual language of the face. Better mechanization of this universal visual language will enable amazing entertainment applications from toys to comforting companions for the elderly (see Figure 5.2). Even in a military scenario, wherein a robot must communicate swiftly with human soldiers, the power of emotive communications could be useful and should not be underestimated.

Several academic laboratories around the world are rapidly progressing with paralinguistic expression robotics. Some use computer-generated faces on screen displays, while others use cartoonish or mechanistic faces as well as replicating the human appearance in lifelike virtual robots. Peter Plantech describes and discusses the development and implications of virtual robots in his (2003) book *Virtual Humans: A Build-It-Yourself Kit, Complete With Software and Step-By-Step Instructions*. Of the few who try to reproduce the human likeness realistically, most are in Japan. In Japan, the researcher Akasawa has built a realistic smiling head using nitinol wires as actuators. Given the temperature dependence of nitinol, this robotic smile takes seconds to cycle through, inhibiting the efficacy of the robot in real world applications. Fumio Hara and a number of animatronics companies have used hydraulic actuators to affect facial expressions; however, hydraulic actuators require a cumbersome compressor and significant level of power, preventing mobile and nontethered applications. The robots that were developed by Hiroshi Ishiguro and actuated pneumatically also suffer from similar drawbacks. Electrically driven and battery operated robot faces allow them to be more mobile and can be made to display realistic facial expressions.

In the MIT AI lab, the sociable infant robot Kismet relies on highly simplified cues of facial expressions without disguising its mechanical identity. The interactive and

Figure 5.2. A futuristic vision of using a humanlike robot caring for the elderly. Such robots may allow aging persons to continue living in their own home while being monitored and helped in case of emergency.

sociable nature of this intelligent robot serves to endear Kismet to observers; it also demonstrates the viability of applications with less naturalistic stylizations of the human figure, particularly if the AI is well done. The machine identity of Kismet highlights a salient robot design principle: The efficacy of the aesthetic identity of a robot relates directly to the aesthetic qualities and limitations of available technology. As biomimetic movement and control catch up with biomimetic visual aesthetics, people may accept robots with less fear (see Chapter 7). Proof comes from biomimetic robots in theme parks, which are generally accepted with great public enthusiasm, perhaps thanks to the painstaking animation by human artists. Such an animation process could enhance other robot applications as well.

There is much to be said for striving to move away from stereotypical machine identities for robots. The human facial system communicates significant quantities of information very quickly, and robots will be greatly improved by emulating the subtle extremes of this system. Because our senses are so finely tuned to healthy movements generated by highly evolved biological facial tissues, it has proven challenging to achieve full aesthetic biomimetics using conventional engineering practices. Exploring the aesthetics of robots may also require artistic practices to create robots. Cognitive and aesthetic design principles will be employed increasingly as we have more face to face encounters with robots, and as robots further proliferate in the marketplace. The success of such encounters will depend substantially on the aesthetics of given robots and their usefulness to users.

THE ART OF MAKING HUMANLIKE ROBOTS

Presently, even the most realistic robots may seem somewhat dead, because, in many ways they are. They are only partly aware, and they shut down instead of going to sleep. They can also break. These flaws in their humanlike appearance can remind us of our own mortality. They also suggest the act of impersonating humans, conveying the threat of an imposter. However, if we remove these flaws to make them friendly, attractive, and seemingly alive, then the level of realism may not matter. Ultimately, good design may help us to make robots lovable and a partner to our human family. More freely exploring the full range of robot aesthetics will certainly accelerate the evolution of humanlike robot design. Moreover, the expanded exploration promises to help us better understand human social perception, interaction, and cognition.

In tandem with the technology, the aesthetics of humanlike robots are evolving at a rapid pace that promises exciting new works of art and entertainment. If this technology actually achieves sentience, as many trends and pundits have predicted, then the role of art will shift to that of artist, just as the role of invention shifts to that of inventor. In this concatenation of works, the artist exercises particular focus on the creative process of the artist/inventor as an extension of evolutionary physics. In the systemic aspects of arts and technology, the aesthetic applications of technology fundamentally change the direction of the technology and its adoption by users and markets. Thus, the aesthetic maturity of humanlike robotics will be instrumental in the proliferation of the technology into our lives. This is not to downplay the possible negative effects of humanlike robots on people and the world, and in the course of realizing humanlike robots, we must consider the consequences of proceeding in this path of inquiry.

Aesthetics of Humanlike Robots

Aesthetically appealing bio-inspired robots do not need to look exactly like us, or precisely mimic the human bio-system. Some biorobotic applications will resemble the biological systems that inspire them, while others will spin the principles derived from nature in new, unexpected directions. In the course of our lives, whether human-inspired robots look realistically human or merely benefit from using bio-derived principles, they will pose major challenges to human identity. Humanlike conversational robots will not only push the aesthetics of robots forward but accelerate the engineered intelligence of the robots – the most complex branch of humanoid robotics – because such intelligence will produce ever more rapid innovation of still smarter intelligence technology.

Human identity will not hold steady in the course of the evolution of humanlike robots. Even as robots increasingly resemble us via biologically inspired engineering practices, we will come to resemble them via the technological augmentation of our physical selves. From all directions, nature, technology, and humanity may merge. The outcome of such convergence would seem like sheer reverie if it were not for the overwhelming amount of real progress in both arenas, generating increasingly useful applications. The complexity of mass-produced products will rise. Instead of a very few actuators, we will likely be able to use tens of thousands of inexpensive discretely controlled actuators. Likewise with sensors, the critical element for interacting with the world; progress will revolutionize the performance of robots. Lack of dense sensors is one of the prominent reasons why robots are still dumb. Consider also the ongoing rapidly emerging improvements in processor speed, efficiency, and power; in memory; in self-improving software; and in AI. The marketing of such technologies are also expanding rapidly. By many metrics, in coming decades machines promise to be as smart as humans.

The charge inherited by the developers of biomimetic applications is to ensure that their robots emerge as ethical; the intelligence of these robots must be inspired by what is good in human nature, with conscious avoidance of the shortcomings (and evil) inherent in the biological underpinnings of the human mind.

Introduction into Our Lives

Certainly we think of virtual reality similar to how we feel about the worlds that are woven in video games (see Figure 5.3). However, the world of theme parks, which brings the artificial world of movie animation to life, also represents powerful forms of artificial realities that become stronger in their future implications, especially if we consider robotics as physically embodied computer animation. Robotics is increasing its efficacy at a startling rate, and the constituent components (sensors, motors, and advanced materials) are plummeting in price. In the process, the basic technologies and their integration are becoming easier to reformulate into new products and to distribute into the marketplace.

In recent years, there has been a great shift in the economy to a largely theme-based marketplace. Restaurants now sell the magic of a highly exotic, other worldly environment as much as they sell food. For example Pappadeaux, La Madeleine, and Olive Garden are all successful restaurant chains in the United States. Shopping malls and retailers have focused on the disorienting and dazzling theming of retail spaces to entice customers. We live immersed in virtual worlds of sitcoms and surreal marketing dream worlds spun in amazing 30-s commercial breaks. Perhaps unconsciously, we like to live in dreams, but we certainly reward the companies that immerse us in them and avoid those that cannot provide the refined themed experience.

In the fine arts, one can also find increasing shifts toward immersion in themes as typified by the works of such artists as Mike Kelley, John Baldassari, Cindy Sherman, and Paul McCarthy. Often art presages the cultural mainstream, and we can anticipate seeing derivations of these aesthetic explorations begin to surround us in both virtual and physically embodied forms. These two robotic forms and the demand for themed commercial immersion will inevitably dovetail in wondrous immersive environments.

Figure 5.3. A video games arcade in Japan, where the users are playing simulation games that incorporate virtual reality. Photo by Adi Marom.

In this increasingly theme-park world, our world is becoming filled with phony representations, including plants, trees, streams, songs of birds, and other soothing voices in the background. A striking difference between the old fiction of theme parks and the new fiction of immersive theming is that the new fiction is becoming ever more "real." The science fiction of intelligent robots written by real humans is now supplanted by real robots that present a fictional representation of humans.

VIRTUAL AND ACTUAL MIMICKING OF HUMANS IN ROBOTS

As we emulate nature, we actually embody it ever more within us. It is a race to see if we can preserve more nature than we destroy and if we can save ourselves. We are scanning genomes, unlocking proteomes, scanning the human brain at ever finer spatial and temporal resolutions, and embodying all these things in ever more refined mathematical models and computer simulations. In all these processes we are extensions of evolving nature, accelerations in the pace of evolution. Our waking dreams that are the trends of the immersive modern culture can be thought of as extensions of this natural evolution. They represent the evolution of the meme. It is in this kind of spirit that we can turn the greed of the consumer culture into something a bit more benevolent. Redemption may come essentially by making the themed elements real again, remaking reality, adding depth rather than compression to our works.

In time, robots may become more intelligent than humans, more compassionate, "more human than human," in a slogan from *Blade Runner*, based on ideas of Dick (1968). Additionally, we may enhance ourselves by neural interfaces, by genetic augmentation of our own brain structure/size, and possibly by the very digitization of our minds. Kurzweil, Merkle, Moravec, and other "transhumanists" predict these changes will hit within 30 years. Whether or not these trends bear out fully, our machines will likely continue to become more biological (bio-inspired), and we will likely become more effectively interfaced with our machines. However, if the bolder trends do bear out then we will fuse with our machines to form a rapidly evolving VALIS (vast active living intelligence system), as anticipated by Dick in 1981. Already, we see that enhanced computational tools have unleashed an era of unprecedented imaging and modeling of natural processes.

With increasing levels of nuance in the simulation of life and nature, computer imagery has begun to blossom into a diverse evolutionary tree of arts, scientific models, and technologies. Entertainment animation, arguably the most salient of such activities, imitates life in film, television, and video games. Yet many of the algorithms used in entertainment animation also serve in flight, military, and surgical simulators. Surgical simulations, in particular, push the physical verisimilitude far beyond that of entertainment animation, integrating considerably more human biological science. This integration includes complex formulae regarding the nonlinear physics of soft tissues and bodily fluids. That said, such simulations do not yet adequately represent the complexity of human tissues; and the science is presently struggling to provide better models and algorithms for these simulations.

Other sectors of science are simulating life in ways that are still more abstract. Artificial life computationally simulates organisms in ecosystems and can simulate the behaviors of these systems over large time scales to investigate evolutionary processes. Further, bio-informatics simulates life in a still more abstract way by attempting to visualize the systems and processes at the molecular and physical foundations of life. Proteomics, which is an extension of the field of bio-informatics, attempts to model the systemic interactions of DNA, RNA, and proteins as the basis of all living processes.

Bio-inspired robotics and materials do not just mimic the external appearance of organisms and biological materials and processes; they model the physical principles that enable living creatures to have marvelous properties and capabilities and then transfer this knowledge into engineered materials and machines. The result is reverse-engineered biology implemented in machines and materials that effectively and truly emulate biology. This multidisciplinary practice is just getting started and is largely supported by better computer simulations and other computational tools. Computational neuroscience is attempting to model human and animal nervous systems as well as the processes of consciousness itself. AI also loosely attempts to model intelligence, usually not by algorithms that are biologically inspired but by various humanlike behaviors (pattern matching, language processing, facial perception, etc.).

Bio-inspired robotics and computational imaging are messing aggressively with what it means to be human. The question of what is simulated and what is real is increasingly becoming fuzzy. Some of the forms of modeling, while admittedly not obviously photorealistic, are actually giving machines the ability to act more like animals and people. Many of these techniques (such as behavior-based AI in gaming) are bearing fruit today. Other modeling techniques (such as computational neuroscience) may not really bear fruit in improving the behavior of realistic humanlike agents for many years to come. However, it is the trend that is remarkable, a trend that is booming, a trend that implies that simulations will increasingly be not just skin deep but will become deeply real.

CRAFTSMANSHIP AND VIRTUOSITY

There are numerous fronts for craftsmanship and virtuosity in rendering animated characters and environments; these largely depend on the medium of choice. In software, the virtuosity may involve the invention of novel algorithms or architectures. In robotic technology, the mechanical, electronics, and fabrication materials must be mastered to produce a satisfactory result. For an emulated personality, a great deal of software integration and interface design must be undertaken. Additional to the technical masteries, there must be a great deal of artistic mastery, especially for depicting extremely realistic or cute characters. In the animation community, a cute and expressive cartoon is often considered to be just as difficult as a realistic face or form, arguably more so because of the strange process of abstracting and incorporating what is human into a nonliteral interpretation.

Whether real or cartoonish in depiction, the aesthetic process of building artificial realities and characters is presently largely determined by the artist's intuition and not scientific principle. It would be desirable to understand better the techniques of the artist from a cognitive perspective and as a series of formal rule-based steps, to transfer the techniques from the black box into engineering principles. Ultimately, such principles could be modeled using programming and parametric models, so as to automate the design process. Initially, this modeling could concern the mechanical aesthetics but, in due time, could include the creating of intelligent personalities for robots and agents as well. Such tools would allow characters and realities to be generated much more quickly and efficiently with a more reliable aesthetic.

In the meantime, the vigorous debate will doubtlessly continue regarding what aesthetics are possible and best suited for given applications. Such debate is healthy and propels the transition of intuition toward design principle, provided the proposed principles are properly tested in controlled (cognitive science) experiments. Such experiments should anchor the theories to neural, evolutionary, and cognitive processes. Of course, culture is extremely complex, and cognitive science is much more basic and not fully capable of formalizing or emulating the full tapestry of the human culture. A close association of art, engineering, and cognitive science will produce increasingly useful applications. A standing dogma regarding the limits of virtuosity is that one cannot create an appealing, nearly realistic humanlike character. Dubbed the theory of the "Uncanny Valley," the theory holds that cartoons and perfectly realistic depictions are appealing, but the less than perfect ends up in the "valley," or unappealing region (see Chapter 7). However, nobody has rigorously tested the theory. Instead, proponents merely cite movies such as *The Polar Express*, saying that such depictions are unappealing and so prove the theory. Contrary to this citation, however, very abstract cartoons can be scary and grotesque (such as a cute but decapitated cartoon character), and cosmetically atypical real humans can be startling or uncanny to behold. The near-real "David" statue of Michelangelo is generally regarded as highly appealing, and the motion of the Sony humanoid Qrio is also regarded as very appealing.

Any level of realism can be appealing (or disturbing), depending on the quality of the aesthetic design. However, to create appealing robots requires true virtuosity and the realization of levels of craft well beyond that of previous robots and anthropomorphic depictions. The neuroscience of facial perception shows that people are much more favorably responsive to cute, symmetrical faces within a limited aesthetic domain, especially if a given face has good skin, healthy-looking eyes, and sweet facial expressions that are socially responsive. These are the aesthetic challenges for realistic robots and animated agents.

BIO-INSPIRED VERSUS BIOMIMETIC

Although realistic humanlike depictions can be useful and interesting, certainly not all robots need to be realistic or lifelike. There is no single right answer regarding aesthetic choice. Rather, there is a wide plateau of aesthetic options to be explored artistically. When it comes to aesthetic possibilities, the field is like the "Wild West." As the tools for rapid custom design of robots and agents (authoring applications) arrive in the hands of artists, there will be an explosion of intelligent, interactive narrative as an art form that will lead in surprising and unpredictable directions. The characters could look like animals or fantastic creatures, cartoons or mechanical men. They could look like celebrity portraits or celebrity caricatures.

Even at this time, there are cases where abstraction can be better employed than realistic portrayals. For instance, a cartoon character from television and movies might be much more effective as an automated teacher for small children. An exaggeratedly cute baby face (with unrealistic features, including an extremely large forehead, cheeks, and eyes; a very tiny nose, and small suckling lips) can inspire the nurturing instincts of children and adults alike and cause adults to forgive technological deficiencies in the robot or agent. Since abstracted depictions simplify the human reaction and may receive a more consistently favorable reaction, it may be wise to choose abstractions early in the development of robots and agents to establish favorable reactions to robots in general.

In many ways, art and fiction routinely twist, challenge, and transform human identity. Therein, our culture – our "software" (memes) – is reinvented, ostensibly propelled forward as an extension of evolution. Figurative arts redefine the human visual identity, while the robots of science fiction redefine the conceptual boundaries of the human being. As an artistic and narrative medium, however, the actual AI-driven robot both redefines our visual appearance and the conceptual framework of the human. Because robots may be programmed to act much smarter than they actually are, robots are, in effect, a new fictional medium, physically embodied science fiction. By creating these robots, we are able to confront issues of the future before the issues and the future spin out of control.

In other ways, we know we are totally independent of our faces. We are mental beings, intellectuals; our ability to transcend our biological biases certainly distinguishes us as human. In our ideals, we do not judge based on the attractiveness of a face. We like to think we can transcend our biological tendencies to favor the attractive and to exclude the physically unattractive, and even our tendency to fear things based on their (distorted) human appearance (i.e., puppets or clowns).

AESTHETIC BIOMIMESIS AND ENTERTAINMENT

Of all the domains of bio-inspired engineering, aesthetic biomimesis is the most vital for entertainment applications. Engagement of an audience is the essential purpose of entertainment. Although this purpose can be greatly enhanced by advanced technology, such as bipedal locomotion and AI, ultimately it is lifelike aesthetics and the ability to communicate that make entertainment effective. Since prehistory, the fine arts have mimicked the aesthetic and communicative aspects of organisms. Examples abound: Michelangelo, Da Vinci, and Disney are all best known for portrayal of living creatures. Although artists do not tend to formalize their powerful methods of biomimesis, in the late twentieth century cognitive psychologists began to consider scientifically the related questions, mostly in the field of dubbed paralinguistics, with promising preliminary results.

Efforts to codify human facial communication resulted in Ekman and Friesen's famed Facial Action Coding System (FACS). Body language was also studied and further increases the sociable applications of robotics. This and future work that relate physiological expressions to increasingly subtle cognitive states will be extremely useful when automated into sociable, interactive, entertainment robots. The practical application of the science of human communications and visual cognition can be considered aesthetics engineering. In spite of such engineering potential, for the foreseeable future, less robustly formalized aesthetics can be expected to be the norm in entertainment robotics.

The heuristics of artistic talent can be considered a black box problem solver for tasks of aesthetic biomimetics and can thus be used in the same manner as a neural net (itself a biomimetic technology). The trained artist simply extends the natural human faculty for communication into external tools and materials (see Figure 5.4 for an example of aesthetic biomimetics). This black box approach has proven essential in getting sociable and communications biomimetic applications to market, as demonstrated in the design of biomimetic toys, video games, and other sundry devices. In addition to artists' static design work (such as sculpture, painted surfaces, etc.), artists are also integral in designing movement in biomimetic applications.

Stan Winston Studios, Walt Disney Imagineering, and Jim Henson Creature Shop, among dozens of other animatronics (themed animation robots) shops, already apply nonverbal facial and body language regularly in robotically animated narrative arts. Animatronics, computer animation, and video games generally employ human animators to design movement, particularly when the movement is narrative or

Figure 5.4. Kelly, a sculpted portrait that demonstrates aesthetic biomimetics, provides a concept for enhancing human–robot interaction.

character-driven. Animatronics, with its "down-and-dirty" approach that emphasizes commercially presentable results, has achieved the highest degree of success in history in mechanical aesthetic biomimetics, evidenced in copious feature films such as Spielberg's *AI* and Disney's *Inspector Gadget*.

Although the artist is essential now, as science and engineering progressively unlock nature's secrets, aesthetic design will be increasingly automated. At the point that machines can truly comprehend and enact narrative, though, they will surely do vastly more than animate movies. They will write, direct, purchase, and watch them.

UNCANNY (UNHEIMLICH) AND UN-KOSHER HUMANLIKE ROBOTS

In 1919, Freud wrote "Das Unheimliche" ("The Uncanny"), inspired by the 1906 essay of Ernst Jentsch, *On the Psychology of the Uncanny* and by the gothic sensibilities of the literary culture of that era. Freud's essay related the uncanny to the anxiety that results from real and imagined surreal circumstances that can be familiar and strange at the same time. In the Freudian *unheimlich is* a surreal separation of the real and the artificial and in particular the artificial as an unpredictable and frightening entity. This concept clearly corresponds to a violation of the standard classifications of people versus objects, and there is a fuzzy boundary between the not quite right – "not kosher" – and "the uncanny."

It may be possible that a humanlike robot can be made to be lovable and not surreal or eerie. This may be simply a matter of skillful design, careful technique, and social acclimation to intelligent humanlike robots. Until these criteria are met, however, the viewer's feeling of the uncanny may plague people's encounters with humanlike robots.

Uncanny, which etymologically means "not known, not safe, or not comfortable," is defined by the dictionary as "seeming to have a supernatural character or origin: eerie, mysterious" and "so keen and perceptive as to seem preternatural." Additionally, uncanny is commonly used to describe a reproduction that is so close to the original that it is startling.

The theory *bukimi no tani*, or "uncanny valley," first posited by roboticist Masahiro Mori in 1970, contends that although cartoonish depictions of humans are appealing (as are perfectly realistic depictions), there exists a "no man's land" between the two that is inherently disturbing. Today, the uncanny valley theory remains one of the most commonly cited design principles in humanoid and humanlike robotics. Although some studies show that people are sensitive to humanlike facial depictions, other studies show that the valley, or dip, into the negative reaction of near realism may be the effect of poor design. This work indicates that, if the aesthetic is right, any level of realism or abstraction can be appealing. If so, then avoiding or creating an uncanny effect depends on the quality of the aesthetic design, regardless of the level of realism.

People in the eighteenth and nineteenth centuries were particularly fascinated by the themes of the uncanny, which inspired the great gothic novels, including Shelley's pertinent classic *Frankenstein*. This is one of the great works to address the uncanny, and it exhibits special (if not downright uncanny) relevance to many issues that are raised when creating sociable robots. In this novel, humans reject the monster solely due to a grotesquely distorted appearance. Driven mad by the pain of abandonment, the monster turns its amazing talents away from creativity to a spree of annihilation. The mythic rejection of this character mirrors Mori's fear that a realistic robot would be loathed as if it were a walking corpse. Yet, if we can scientifically and technologically unravel the mysteries of the uncanny effect, there may be hope of sparing our future robotic progeny from such painful abandonment, even while empowering them with the full bandwidth of nonverbal expressivity that only a realistic face may provide.

Currently, a great deal of robotics, visual perception, and neuroscience research is dedicated to understanding better the uncanny effects in humanlike robots. As a result, great improvements are being made in the neuroscience of human perception of humanlike robots and the parameters of what makes such robots disturbing and appealing. Although the outcome of this work may yet prove that extremely humanlike robots are disturbing, the evidence to date suggests the opposite.

The specific aesthetic considerations for designing a robot appear to be defined by the nature of the human nervous system and its evolutionary history. Existing branches of science concerning human facial, social, and object perception may provide a basic template for what robot designs will be kosher and which will not.

NEUROSCIENCE AND OUR EVOLUTIONARY HERITAGE

The human mind is highly attuned to the perception of the healthy, sentient human being, and in particular, the human face. A great deal of evidence has shown that we are much more favorably responsive to faces that follow certain biologically determined rules that have arisen through natural selection during the evolution of our species. In particular, our attraction is strongly biased in favor of neonate features (high forehead, large eyes and cheeks, small nose, and small suckling lips) and for features of symmetrical, highly sexual maturity.

Signs of unhealthy skin and hair will universally repel a person, as will atypically asymmetrical features or atypical disproportion of features, outside the schemes of exaggerated femininity, neoteny, masculinity, or senescence. Within those schemes, disproportionate features are characteristically appealing. This is exhibited in cartoon films and illustrations.

In 1971, Ekman and Friesen established the foundation of paralinguistic values of facial expressions in the course of conversations. These expressions may be used to put a person interacting with a person (or robot) at ease and establish a normal flow of conversation and attention. It remains to be seen whether other aesthetics may be acceptable in robots. Also, the knowledge that one is interacting with a robot and not a human may alone cause some discomfort in people – some people more so than others.

Experience shows that people get quite used to interacting with robots and come to feel quite at ease. This, of course, needs to be backed up with formal studies. One thing is certain: The human nervous system has a highly tuned, innate aesthetic structure that will only tolerate a narrow range of aesthetic representation in art, animation, and (we feel certain) robots. Fiction, art, and myth exist as highly evolved neurocognitive phenomena. The fictional art of robots must function as a sort of "paint" on the "canvas" of the human nervous system.

Long before robots are truly sentient, deft artists will create the illusion of sentience. Long before a robot can truly feel love, the robot will act convincingly in love with its human companion. It is important to fill the time between faux-bots and strong AI with aggressive study of human and robot social intelligence. The human face and the emulated human face on the machine can serve as excellent tools for ferreting out the nuances of social cognition, formalizing them, and then emulating them in machines that grow ever more socially cognizant and caring.

TWENTY-FIRST CENTURY ART

Bio-robotics and AI-driven computer animation promise to transform the face of entertainment and art in ways no previous art medium has. We are talking about art that is sentient. Such art can evolve in several directions. As entertainment or traditional character-driven narrative, such robots will be animated characters brought to life, and they may have their own stories, feelings, and motives.

Robots that incorporate computer vision are equipped with the ability to see faces, gestures, and objects. Thus, the robot can make eye contact, recognize an individual by face, and respond in a friendly way. Robot systems also use speech recognition and a synthesized voice while lip-syncing to the voice in order to appear as if holding a natural conversation. Animating a robot is similar to traditional computer animation in that it is generated by motion capture or by an artist. Once the robot animation is generated by an artist, its system needs to play it back in a way that is modulated to a specific context: conversation based on either visual stimuli or the mood or memory of the robot.

The personality of the robot is also similar to conventional character design in that it is prepared in a form very similar to a movie script. The difference is that the writing defines the general conversational patterns of the robot, its beliefs, its history, and its story. Additionally, the flow of conversation is devised in general, with numerous conversational states (i.e., general greeting mode, storytelling mode, tutoring mode, etc.).

Achievable approaches allow faking a great deal while having small "smarts," which can be based on some extant systems and some experimental inputs. These smarts may incrementally be enlarged with time and system use. The best methods will involve the elaborate modeling of many parts of the human mind, simulating a biographical memory, understanding the stories and knowledge it is told, stashing these, and recalling them in an appropriate context. Such systems will grow more effective in the coming years, and their related techniques will enable, say, a television character, to become part of one's family and to weave character narratives into its response to the family. In the process the developed technology and artistry will define a new art-form. The manufacturing technology and the market performance of resulting consumer products will drive the evolution of such entertainment character robots.

In addition to the expected development, these robots will find their way into extreme conceptual art. Already, several radical artists are devising animatronic figures that push beyond the boundaries of good taste and challenge human sensibilities. For example, McCarthy uses robotic hardware in images of abject moral bankruptcy. The quality of the animatronics, generated by Kindlon (formerly of Rick Baker and Stan Winston's studios) is exquisite, inspiring Disney Imagineering to inquire into the use of the robots for their parks. For example, Ken Feingold used chat-bot software to create endless recursive conversations between sculpted heads. The mechanization of the human form reminds one of machine-like obsessive loops that humans experience, evoking that same sense of helplessness of individuals trapped in a kind of physical limbo, never to escape. Although the casual observer may find such artistry trivial, offensive, or frightening, these works are raising numerous issues regarding the essence of humanity. Here we see the darker aspects of the mechanization of humanity. Alternately, we also see works of art that are coming to life in ways never before possible.

Many questions arise in relation to these objectives: Just what is art? How can these machines function as art? What will they become? What aesthetic, social, cognitive, and psychological issues writhe beneath the rock of realistic human emulation (especially given the limitations of the present technology)? Will they be inherently repulsive? What are the systems of human cognition (a quick sketch), and how may they be approximated currently? How can social interaction between robots and humans be constrained so that it is maximally effective, given the extreme limits of the technology? As a corollary to this last question, what can be faked – or dodged – in such interactions?

Under the artist's objectives, robots may be shaped into aesthetically expressive objects, but the artist generally seeks more than this. Beyond mere slick designs, the artist will seek a conceptual or psychological impact, one of some novelty and meaning. With robotics, this motive will result in a branching diversity of robotic forms and functions. With this in mind, one may anticipate seeing unleashed in this century a strange and wondrous panoply of images, living, walking, and thinking, images that shift through time, that learn, and that love. Some of these machines will inspire love. Others will be grotesque or even repulsive. Some may inspire fear, while others will become part of our families, perhaps beloved friends to our children. However, for this last thing to happen, robots must earn our trust.

Several obvious potential applications for humanlike robots threaten to spoil our trust. One, the Pygmalion phenomenon, in which we feel physical and romantic love with our robots, is almost certainly imminent. Here, however, we will find the threat of our own dark sides, which will almost certainly be pinned on the technology itself. Before the robotic cathouses and pleasure droids become widely known in the world, let us hope that kindly robots are known first. The outcome of the generation of artistic derivations of technology will flower like a rain forest in diversity. The outcome cannot be predicted.

SUMMARY

The act of developing humanlike robots may appear trivial at first, but upon more profound reflection, one finds that it is a step in the history of the evolution of life that promises to radically change human identity. At this early stage of robotics development, we may use this reflection to examine what it means to be human and also to try to learn what is canonically human. These explorations using robots may be philosophical, scientific, or artistic. Of course, scientific exploration will be more formally rigorous. However, even the informal and artistic explorations effect change in the mindscape and in our cultural information landscape. They can forever change our expectations of what robots are and should be, and what humans are and what we may become in the future.

BIBLIOGRAPHY

Books and Articles

Asimov, I., *Runaround* (originally published in 1942), reprinted in *I Robot*, (1942), pp. 33–51.

Asimov, I., *I Robot* (a collection of short stories originally published between 1940 and 1950), Grafton Books, London, (1968).

Bar-Cohen, Y. (Ed.), *Biomimetics – Biologically Inspired Technologies*, CRC Press, Boca Raton, FL, (November, 2005).

Bar-Cohen, Y., and C. Breazeal (Eds.), *Biologically-Inspired Intelligent Robots*, SPIE Press, Bellingham, Washington, Vol. PM122, (2003).

Birdwhistle, R., *Kinesics and context: Essays on body motion and communication*, University of Pennsylvania Press, Philadelphia, PA, (1970).

Borges, J. L., "Narrative Art and Magic" Translated by N. T. Giovanni in *Borges: A Reader*, E. Rodriguez and A. Reed (Eds.), Dutton Press, New York, (1981).

Breazeal, C., *Designing Sociable Robots*. MIT Press, Cambridge, MA, (2002).

Darwin, C., *The expression of the emotions in man and animals*, D. Appleton and Company New York, (1913).

Dawkins, R., "Memes: the new replicators," Chapter 11, in *The Selfish Gene*, Oxford University, Press, Oxford, 1976.

Dick, P. K., *Do Androids Dream of Electric Sheep?* Rapp & Whiting Press, (1968).

Dick, P. K., *VALIS*, Bantam Press, New York, 1981.

Ekman, P., "The argument and evidence about universals in facial expressions of emotion," *Handbook of Psychophysiology*, H. Wagner and A. Manstead (Eds.), John Wiley, London, (1989).

Ekman, P., W.V. and Friesen, "Constants across cultures in face and emotion," *Journal of Personality and Social Psychology*, vol. 17, (1971) pp. 124–129.

Etcoff, N., *Survival of the prettiest – The Science of Beauty*, Anchor Books, (July 2000).

Fancher H., Screenplay, *Blade Runner* http://www.trussel.com/f_blade.htm, (1980).

Fong, T., I. Nourbakhsh, and K. Dautenhahn, "A survey of socially interactive robots," *Robotics and Autonomous Systems*, Vol. 42, (2003), pp. 143–166.

Frauenfelder, M., "Baby Robots Grow Up. Emotionally expressive artificial companions prove that functional adaptability outweighs technology perks," *International Design Magazine*, Vol. 49; Part 3, (May, 2002), pages 52–53.

Freud, S., *The standard edition of the complete psychological works of Sigmund Freud*,Translated by James Strachey, W. W. Norton & Company, (January 2000).

Hanson D, D. Rus, S. Canvin, and G. Schmierer, "Biologically inspired robotic applications," Chapter 10, *Biologically inspired intelligent robotic*s, Y. Bar-Cohen and C. Breazeal (Eds): SPIE Press, Bellingham, Washington, Vol. PM122, (2003), pp. 285–350.

Hanson, D., "Expanding the design domain of humanoid robots," *Proceeding of the ICCS Cognitive Science Conference, Special Session on Android Science*, Vancouver, Canada, (2006).

Hanson, D., A. Olney, M. Zielke, and A. Pereira., "Upending the uncanny valley," *Proceedings of the AAAI conference*, (2005).

Hara, F., H. Kobayashi, F. Iida, and M. Tabata, "Personality characterization of animate face robot through interactive communication with human," *First International Workshop in Humanoid and Human Friendly Robotics, International Advanced Robotics Program (IARP)*, Tsukuba, Japan, (October 26–27, 1998), pp. 1–10.

Hofstadter, Douglas R. *Gödel, Escher, Bach, and Eternal Golden Braid*, Basic Books. (1999), p. 10.

Ishiguro, H. "Android Science, Toward a New Interdisciplinary Framework." *Proceedings from Android Science*, Stressa, Italy, (25-26 July 2005), pp. 1–6.

Kelley, M. *Mike Kelley: The Uncanny (Art Catalogue)*, Walther Konig (May 2, 2004).

Kindlon, D., Personal conversations with the co-author David Hanson, Hanson Robotics (2007).

Kurzweil, R, *The Age of the Spiritual Machines: When Computers Exceed Human Intelligence*, Viking Press, New York, (1999).

Kurzweil, R, *The Singularity Is Near: When Humans Transcend Biology*, Viking Press, New York, (2005).

Langton, C. G., *Artificial Life: An Overview*, MIT Press, Boston, MS (1995).

Levenson, R., P. Ekman, and W. Friesen. "Voluntary facial action generates emotionspecific autonomic nervous system activity," *Psychophysiology*, Vol. 27, No. 4, (1990), pp. 363–383.

Masschelein, A. "A homeless concept, shapes of the uncanny in twentieth-century theory and culture," *Image & Narrative*, Online Magazine of the Visual Narrative – No. 5, (January 2003).

Menzel, P., and F. D'Aluisio, *Robo sapiens: Evolution of a new species*, MIT Press, Boston, (2000).

Mori M., *The Buddha in the Robot: A Robot Engineer's Thoughts on Science & Religion*, Tuttle Publishing, (1981).

Mori, M., "The uncanny valley," *Energy*, vol. 7, No. 4. Translated from Japanese to English by K. F. MacDorman and T. Minato, (1970), pp. 33–35

O'Toole, A. J., D. A. Roark, and H. Abdi, "Recognition of moving faces: A psychological and neural framework," *Trends in Cognitive Sciences*, vol. 6, (2002), pp. 261–266.

Plantec, P. M., and R. Kurzwell (Foreword), *Virtual Humans: A Build-It-Yourself Kit, Complete With Software and Step-By-Step Instructions*, AMACOM/American Management Association, (2003).

Premack, D. G., and G. Woodruff, "Does the chimpanzee have a theory of mind?" *Behavioral and Brain Sciences*, Vol. 1, No. 4, (1978), pp. 515–526.

Rees, G., G. Kreiman, and C. Koch, "Neural correlates of consciousness in humans," *Nature Reviews Neuroscience*, vol. 3, No. 4, (2002), pp. 261–270.

Reichardt, J., *Robots: Fact, fiction, and prediction*, Viking Press, Middlesex, England (1978).

Shelley, M., *Frankenstein*, publishers: Lackington, Hughes, Harding, Mavor & Jones, (1818).

Thorrisson, K. R., "Mind model for multimodal communicative creatures and humanoid," *Applied Artificial Intelligence*, Vol. 13, (1999), pp. 449–486.

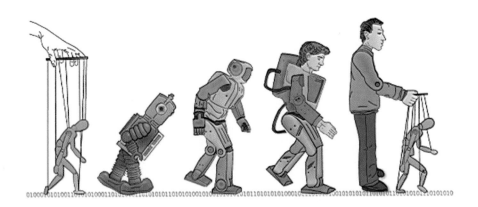

01000101010011010101000110101011101010110101010010101010110110101010100010111010101010100010111010110101010101010110010101010101010110101010

Chapter 6
Trends in Robotics

Humanlike robots may change our workplaces, economy, health care, and many other aspects of our lives. Imagine you are approached by a salesperson that is a robot, possibly like the one shown in Figure 6.1. Now imagine that this robot warmly smiles at you and presents a new electronic gadget. Would this grab your attention? Out of curiosity you may give the robot a chance to talk to you about the product and maybe even sell it to you.

This may be one of the directions in the development of humanlike robots. As they become "smarter," they may be used to solicit customers, show them products, and demonstrate how they work. Such robots could greet you in reception lobbies and guide you to your destination at shopping centers. Or, they could guide you through a museum, delivering educational content and taking your questions to enliven the presentation. In health care, a humanlike robot may keep an elderly person company (see futuristic graphics in Figure 6.2 of Chapter 5) while monitoring his or her health or even possibly perform emergency treatment.

Far in the future, robots with artificial cognition may be developed, but as they become increasingly capable they are expected to raise concerns that are unimaginable in any other product. To understand the significance of having intelligent mobile machines with self-identify let us consider the following scenario. Imagine a situation where a humanlike robot is allowed to participate in a competition such as running a marathon or just taking part in a baking contest. For a simple machine or appliance, there would not be any questions regarding the ownership of the award. However, in the case of a humanlike robot, one may wonder who should receive the award – the winner robot or its owner. A robot with self-identity may demand to receive the

Y. Bar-Cohen, D. Hanson, *The Coming Robot Revolution*, DOI 10.1007/978-0-387-85349-9_6,
© Springer Science+Business Media, LLC 2009

Figure 6.1. A potential humanlike robot salesperson.

award. This opens the issue of the legal rights of such robots. In the larger frame, these rights may resemble human's worker compensation rights and related laws. If in winning a competition or simply providing a service a robot is able to earn money and build wealth, then robots may be able to affect in significant ways how we live our lives. We can certainly anticipate the rise of challenging, unpleasant, or even dangerous situations.

With an increase in their sophistication, humanlike robots may begin to express opinions about our society and possibly even develop political views. As they grow more capable we may begin to wonder if they should be granted freedom of speech or the freedom to demonstrate. Looking at the fabricated illustration of a demonstrating humanlike robot in Figure 6.2 one may feel uncomfortable seeing a robot demonstrating against our institutes, establishments, government, companies, or other parts of our human society. If these scenarios materialize then the possibility that such robots could turn against us may have a higher probability than we like to believe possible. As robots become equipped with artificial cognition their actions and behaviors may not be in agreement with us, or worse, they may become antagonistic with humanity. Additionally, such robots may not be built with the nuanced sense of conscience that humans have evolved over millions of years. They may have little innate conception of right and wrong, or even concern for

Figure 6.2. Potential future protest of robots against humans.

the value of life. Without enabling robots with a fundamental friendliness toward humanity, they may make decisions and actions that are psychotic and devastating. Other issues that would need to be dealt with include the consequences of damaging such a robot, and whether the laws that apply to humans regarding violence and physical injury or even death are going to be relevant in such cases to these machines.

To produce intelligent humanlike robots that are able to have their own thoughts and take independent actions, technology has to incorporate into the machines a stable and conscientious mindset. To understand the significance of this one may think of a baby and our attitude toward that baby as it becomes able to move combined with a great curiosity about the surroundings. Once the baby starts crawling and walking, he or she needs to be constantly in our sight since we never know what the baby may do or be able to reach. Similarly, we will need to monitor the movements of autonomous humanlike robots since initially they may have many errors in their control algorithm and operating system as well as have software and/or hardware that are not reliable. Therefore, we will need to be able to prevent such robots from causing damage to people or property as well as protect them from damaging themselves.

ROBOTS AS SERVANTS OF HUMANKIND

Using robots in the service of humans may evolve in many different directions; it will depend on their capability and the niche applications that they will fill. One may envision a master–slave relation or the use of autonomous robots that are given well-defined tasks with operational restrictions. In this form of dependence, such robots may find a wide range of applications with minimal complications.

With the current state of technology, we are finding humanlike robots that are performing quite dedicated tasks with a limited number of functions and capabilities. As they evolve, one would envision seeing them do more complex tasks that are currently done by humans. Such tasks may include many futuristic applications, examples of which are suggested in this section.

A humanlike robot may be used to feed your dog as needed and take it for a walk in a nearby park while making sure that the dog stays out of trouble. Also, the robot may protect the dog from potential dangers, including safely crossing the street while paying attention to oncoming traffic. Besides cleaning up after the dog, the robot may make sure that the dog behaves by the rules that the owner dictates. Such a scenario of a humanlike robot taking your dog for a walk in the park is illustrated in Figure 6.3. The robot shown in this figure does not exist. It was created graphically here for the sake of providing a visual image of the concept. This same imaginary robot, with a humanlike head with the likeness of Michelangelo's David and a machine-like body, was shown earlier in this chapter and is also shown in the following illustrations.

For serving humans in the form of an education tool, one may consider developing a Robo-Teacher or, depending on improvements in the technology, possibly even a university-level Robo-Professor. Such a robot (see Figure 6.4) can possibly complement the ones that are teaching courses from remote locations via the Internet. Initially, the robot can be designed to teach short courses or provide simple training. Such robots can help dealing with the increased complexity of new instruments, or a humanlike robot

Figure 6.3. Taking your dog for a walk can be an attractive application for humanlike robots.

Figure 6.4. Using a Robo-Teacher or Robo-Instructor can be a great way to communicate information and even teach courses on various topics.

may be used as a temporary instructor to assist with the installation of new hardware and provide initial operation guidance and training. The Robo-Trainer may stay with the customer until he or she is satisfied with the performance and use of the delivered technology. Upon completion of the task, the robot would return to its owners, which may be the factory or the facility that sold the hardware and own the Robo-Trainer. If it is a distant location the robot can be designed to take public transportation as needed in order to return in the safest and least expensive way. Of course, for robots a "red eye" flight or just self-packing in a box and self-shipping would not be considered a hardship.

In case of power failure, one might consider using similar robots as traffic controllers at intersections, replacing the manual task that is performed by policemen and thus removing the risk to humans. In Japan, for example, policeman-shaped figures are used to alert traffic to possible danger at construction sites (see Figure 6.5). In the future, a Robo-Policeman may take this role in an active form and be able to operate over extended time without break (while recharging the required batteries possibly using solar energy). Such a robot can provide more information and support drivers as needed, particularly if changes occur at the specific site due to unforeseen situations.

Figure 6.5. A policeman-shaped figure in a street in Tokyo, Japan, is being used to alert traffic at a construction site. Photo by Adi Marom.

For sport applications one may envision numerous possibilities for humanlike robots. Figure 6.6 shows a tennis player who can serve as your partner either for training or for the joy of playing against a competitor with a selectable level of performance. To promote accelerated enhancement of the capability of humanoid robots for sport applications there is already the challenge of RoboCup, where a team of humanoid robots play a soccer game against a human team of experienced soccer players. Several RoboCup competitions of robots have been held since the early 2000s.

The jobs that humanlike robots will be designed to take will likely include those that involve a great degree of repetition, do not require much skill, and can be handled by a machine. Examples of such jobs may include cleaning floors as well as vacuuming the carpets in buildings and offices. The task can be done at late hours while no humans are around to be in the way of the Robo-Cleaner and thus operating at maximum efficiency and with minimum human factor issues. Humanlike robots will not "call in sick" and

Figure 6.6. A humanlike robot may act as an opponent in tennis and other sports.

will complete the job at a predetermined time within a schedule and with a very predictable performance. They can be observed at all times without any concern of invasion of privacy and can have their task changed on the spot if a higher priority activity needs to be done. Making such robots is still a challenging task for roboticists, but the great benefits to the users and advances in technology may lead to seeing such development.

Making humanlike robots to serve as Robo-Cooks is another possible development. The need to use different ingredients and spices as recommended in various recipes while operating in a kitchen setting is far from a routine predetermined task that can be done by a simple automatic machine. The multi-tasking nature of a cook's job in a kitchen setting can offer a niche application for "smart" humanlike robots. A futuristic illustration of this possibility is shown in Figure 6.7, where a robot is acting as a Robo-Cook in a sushi bar. Although it may look weird when such a Robo-Cook is used for the first time, it would probably become more acceptable if the end result is cheaper, faster, and tastier. Also, the presence of a Robo-Cook may be entertaining and may help attracting customers to a restaurant with such a novel sight and advanced technology.

Humanlike robots in our homes can act as assistants to the elderly, disabled, or patients in rehabilitation and even as nannies. A vision of a futuristic Robo-Nanny taking care of a baby is shown in Figure 6.8. Would you be willing to rely on such robot to take care of your offsprings? The robot may be used to help teaching children various skills, possibly helping them with their homework, and take on other educational roles in a playful manner, which children may find more appealing than listening to a human instructor. They may play games that reinforce the learning of educational material, possibly teach children to play musical instruments, read books to them, teach them various topics in an entertaining way, including how to write, how to swim or ride a bicycle, and many other possibilities. With time, lessons learned and the latest

Figure 6.7. A futuristic illustration of a Robo-Cook (*left*) preparing sushi behind the counter of a sushi bar.

Figure 6.8. Robo-Nanny taking care of a baby. One may wonder how receptive we would be if such a service becomes reliable and inexpensive.

improvements will be introduced into these robots to make them increasingly more effective in dealing with children and serving as better teachers.

And how about humanlike robot tourist guides? Many times, touring companies use a recorded script that describes to tourists what they are seeing while a driver takes them on a tour of various attraction sites. Having a humanlike robot as a tourist guide may offer an attractive application for these robots. The robot can be equipped with a global positioning system (GPS) and an extensive database about the various sites, and it can use its ability to verbally express itself while making body postures as needed to describe such sites in a pleasant and entertaining way. An example of this idea is illustrated in Figure 6.9, where a futuristic robot is shown talking to a tourist on boat. Besides being able to describe the attraction and show humanlike characteristics, the robot may not require great sophistication or extensive capabilities.

In fact, there are many tasks that involve low skill and are compensated with a low pay. Examples of such tasks were described earlier in this chapter. In due time, such jobs will be filled by humanlike robots, and the resulting impact can be enormous.

Figure 6.9. A futuristic view of a Robo-Guide for tourists.

Taking your meal order at a restaurant and delivering drinks and food to your table could be some of these types of jobs. The younger generation would undoubtedly be more receptive to this form of service, especially in fast food restaurants where the order can be taken by a humanlike robot. After delivering the food, the robot may entertain the customers by possibly singing and dancing or maybe providing information of interest, such as a weather forecast and breaking news. To have such a robot at a child's birthday party could become quite the rage; parents may want to hire a Robo-Entertainer or a robot that can clean up, serve food, greet people, and so on. Although initially this might be considered a novelty, eventually it could become common. With time, jobs will be specially tailored for performance by humanlike robots, and there may be a growing market for them as their efficiency increases, their costs drop, and the demand for them expands. At a limited level, we have already started getting used to communicating verbally with a digital phone operator when we call various companies. Today, when you call your bank, airline, or other service providers, the odds of speaking directly to a human are close to zero. Effectively, via our phones, we are talking to a digital operator (i.e., a "smart" machine), answering questions and expressing requests. The content of the communication already involves a broad range of areas, including getting flight arrival and departure information and checking the balance of your bank account or credit card. Submitting an application for a new credit card is done today by talking to a digital operator over the phone.

Robo-actors may allow movie studios in the future to perform movie stunts that are too dangerous for humans to perform. Such filming will provide the needed realistic appearance while avoiding the unnecessary risk to human actors. The use of humanlike figures in the form of dummies is already being done in testing car crashes, where the dummies are instrumented with sensors to determine the forces and potential consequences to human passengers.

Humanlike robots will also probably be used to perform functions that humans struggle with or are too risky. There are many situations in which humans put themselves in danger, including with exposure to extreme temperatures, pressures, radiation, gases, chemicals, possible drowning, being in the presence of falling objects, and operating in mines. One may even envision such robots operating in space, where humans cannot survive direct exposure to the ambient conditions. This can include very low air pressure and absence of sufficient oxygen, as on Mars, extremely high temperatures (as much as $+460°C$) as on Venus, or very cold temperatures (approaching -160 to $-180°C$), as on the moons Europa and Titan. A vision of such participation of humanlike robots in future planetary exploration missions is illustrated in Figure 6.10, where the robot may operate autonomously and independently or serve as an aid to astronauts.

There are many issues that will result from the operation of humanlike robots in workplaces. These issues include the fact that, as such jobs will increasingly be taken by robots, the people who could do the specific low-paying jobs may not have alternatives available to them. Besides the impact on the poor and low skilled employees, this trend may have an impact on future generations of teen-agers who perform such jobs as their first entry into the employment market.

Figure 6.10. One of the envisioned applications of humanlike robots is their operation in areas that are too hazardous for humans, including planetary exploration as well as support of habitation and construction.

Another possible application with potential impact on the job market can be the use of robots to prevent border crossings by illegal aliens who could fill jobs that require minimal skills. Firstly, there would probably be no demand for illegal laborers, since humanlike robots would be expected to become inexpensive to operate and would be "ready to work" with no restrictions other than their hardware or software capability and no worker compensation issues. Secondly, these robots will probably be able to offer improved border monitoring capability to prevent possible entry of illegal workers. As artificial intelligence algorithms become more sophisticated, we may even see humanlike robots taking jobs that require the skills of white-collar employees.

For self-maintenance, some of the existing humanlike robots are already equipped to make sure that they do not run out of power before seeking to recharge their batteries. Other potential options for future self-maintenance capabilities may include taking care of internal failures in case of slight damage, and if needed the robot may be instructed to go on its own to a local service facility for more major repairs. Equipped with GPS and other sophisticated technologies it may be capable of reaching such facilities far from its home base. Efforts to develop the capability of robotic mechanisms to autonomously reach a remote destination are currently underway at many academic and industrial organizations. Some of the related research is being developed to address the DARPA challenge that requires an autonomously operated car be driven in urban area and reach a specified global location without accidents and while obeying the traffic laws along the way. To meet the challenge, cars are being developed with visual recognition systems and algorithms for both collision avoidance and navigation in a dynamic environment such as that of an urban area. Humanlike robots with similar capabilities but operating at lower speeds in more complex terrains will be functioning in our homes, offices, and communities.

Issues of self defense may arise, for example, when a robot walks to a repair facility or is sent to take care of an errand for its owner. It is generally acknowledged that the first person to raise the issues regarding human and robot interaction was Isaac Asimov, who suggested that to have robots act as servants they would have to be allowed to defend themselves as long as they do not hurt humans. What happens, though, if children notice the presence of the robot and decide to challenge it by chasing or throwing objects at it? Such reactions by children may be an issue mostly in the early generations of these robots, while they are still a novelty and may appear weird to the children. Although the robot should definitely avoid interaction with the children, it will have to deal with a potential risk to its hardware. Running out of the area would be the best thing to do while expressing dismay concerning the children's actions. Through image recognition the robot may identify the children and possibly would call their parents to make them aware of their children's behavior. However, if we assume that the only realistic option for the robot is to run away from the area, it may still not be escaping from dangers and risks. The robot may hit a person or a car along its escape route and subject its owner to undesirable liabilities.

In the future, once humanlike robots start getting out of the protective environment of the home there may be a rise in liability issues. If such a robot has a flaw in its program, it can cause injury to a person or damage to property. In such cases, the owner is liable for the consequences even if he or she is not aware of the robot's whereabouts and activity. Even if the owner does not know a priori where the robot is going to be, the owner is still expected to be responsible. One may assume that the liability issue will be similar to a situation where your dog runs out of your property and gets involved in a situation of bodily injury or property damage, with negative financial consequences. As opposed to the liability that is related to a pet or product, here we are dealing with a machine that can be unpredictable, particularly if it has humanlike intellectual capabilities combined with faulty drive hardware or software. The issues become even more complex if the robot has its own volition and decides to act illegally.

PROVIDING ARTIFICIAL ETERNAL LIFE

The desire of humans to be memorialized has a very long history, and there are numerous examples of the various ways people realize this desire. Examples include the ancient Egyptian pharaohs, who built the pyramids in which their bodies were preserved in the form of mummies. Numerous wealthy people hired artists to make statues in their own image. Although still very expensive and quite limited in capability, making an exact replica of ourselves in the form of humanlike robot is increasingly becoming feasible. In the future, one may order a replica in the form of a robot that appears, speaks, reacts, responds, and behaves like its owner and thus preserves its owner's essence in artificial form for "eternity." Periodically, as specified in the will and as long as there are funds allocated to cover the cost, the robot can be upgraded using the latest capability. If in future generations, one may be able to "download or copy" the human memory from the brain, then such information can become part of the robot operating system. Also, in order to simplify the process of emulating the person's behavior, the robot can be programmed to follow the person in his or her daily life and learn the behavior, adapt the voice, learn the postures and gestures, and copy as many characteristics as possible that identify the individual. Eventually, as duplication of intelligence becomes more sophisticated, one may envision the possibility of seeking the wisdom or advice of his or her great-grand-parents by talking to the clone that is in the form of a humanlike robot. And then, one might wonder, will such beings be afforded all the rights and privileges of the individuals they portray? How accurate would such a portrait need to be in order to be considered the cloned person?

In 2005, David Hanson of Hanson Robotics and co-author of this book worked with a team to portray the deceased science fiction author Philip K Dick, as a humanlike robot that is equipped with artificial intelligence and a memory of many pages of Dick's writings. The developed robot was designed to talk with a person (as shown in Figure 6.11) much like conversing with Dick's ghost or the clone of a person who is no longer with us. This robot was presented at the NextFest conference that was held in Chicago that year.

Figure 6.11. Memorial portrait of late science-fiction writer Philip K Dick engaged in conversation with a young fan. Photo courtesy of Philip and Lono Walker, parents of the boy shown in this photo.

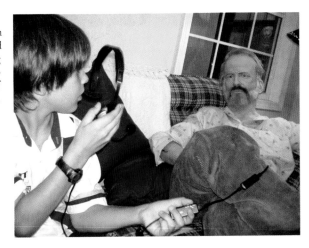

ROBOTS AND HEALTH CARE

Throughout the previous chapters it was mentioned that humanlike robots are finding applications in the fields of medicine and health care. Beyond operating as alternates to health care staff, the technology offers the possibility of making "smart" lifelike prosthetics (see also Chapters 3 and 4). As improvements in making robotic components increase, it will be easier to produce highly effective prosthetics that look very much like natural organs (see Figure 6.12). In addition, health care robots may be used to support medical staff in a variety of tasks, including surgery and treatment of patients during rehabilitation either at the hospital facility or even at a patient's home.

For the elderly, disabled, or patients in rehabilitation, such robots may serve as assistants, providing 24-h, 7 days a week service of monitoring and providing emergency treatment as needed. In case of a serious health-related emergency or deterioration, such robots may be programmed to both perform first-aid treatments and also

Figure 6.12. A prosthetic hand (made by Deka R&D Corp.) under the DARPA program devoted to revolutionizing prosthetics. Photo by Yoseph Bar-Cohen at the 2007 DARPATech that was held in Anaheim, California, in August 2007.

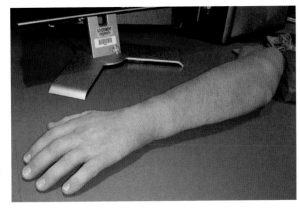

alert health care staff to send help urgently. Thus, we may be able to have the elderly continue to live as independent individuals in their own home rather than needing to have them move to a nursing home when their health deteriorates. Such robots may also serve in nursing homes as part of the "personnel" that takes care of the daily needs of the elderly and thus help reduce the shortage of nurses.

In caring for the elderly who have potential medical risks, robots may be programmed to look for signs of medical problems and provide emergency treatment while notifying the proper urgent care personnel to send help. The robot may perform all the house maintenance tasks, including repairs, cleaning, and personal assistance. It may have in its memory the operation system diagrams of all the house systems and appliances, and it should be able to know how to find what is wrong, how to order replacement parts, and how to fix the problem once the ordered parts arrive. Also, it may play the role of a pleasant companion making conversation and providing current news, possibly even supplying explanations, background, opinions, and answers to questions. A humanlike robot may also serve as a security guard, welcoming strangers that come to the house but able to decide who is allowed to enter and who is an unwelcome intruder. In the case of the latter, the robot may notify law enforcement agents while making sure that the intruder is kept restrained until they arrive.

In Japan, humanlike robots are increasingly being used to assist nurses at hospitals. Some of their tasks include bringing medications and foods to patients as well as keeping them company. An example of a personable robotic nurse in the United States (see also Chapter 2) is one that was developed at the Carnegie Mellon's Medical Robotics and Information Technology Center (MERIT) and tested at the Longwood Gardens nursing home in Pittsburgh. This machine is called the CMU Nursebot (Figure 6.13). It was made to act as a friendly anthropomorphic companion in its interactions with patients, and one of its functions is to remind patients of daily tasks such as taking medicine and going for walks.

Figure 6.13. Carnegie Mellon's Nursebot Pearl at Longwood Gardens in Pittsburgh. Photos courtesy of Sebastian Thrun, Stanford University, California.

In spite of the good robots can do, the elderly are already raising concerns with the future extensive use of robots performing monitoring roles at homes or at medical facilities. This concern was expressed in a survey where the elderly were asked about their reaction to the use of robotic nurses in caring for them. Their major concern is the possibility of being socially isolated from real humans and their being left alone surrounded by machines. You can best imagine how it feels to be surrounded by robots if you think about dealing with digital operators on the telephone. Frequently, one feels trapped and angry when talking to a digital operator and not given the option of directly reaching a human operator. In future these frustrations may be mitigated somewhat by making the machines more flexible and, in effect, more humanlike. The more human-like the machines become, the less we may feel alone and isolated when dealing with them.

Robotic surgery is already a growing area in medicine, due to the incredibly high success rate of minimally invasive operations and the associated shorter recovery time. With success there will be continued growth in the use of this technology, and most likely one may expect the use of humanlike robots someday in the role of surgeons performing operations themselves. However, one may wonder what happens if the results of a surgical procedure are not positive, and how liability laws would apply and how lawsuits seeking compensation would play out in future court cases.

One of the applications that may lead to the most significant advances in the field of humanlike robots is their use in providing sexual services. This prediction is based on the lessons learned from the development of the Internet and the possible market that may be involved with this type of robot. The issues that are associated with the use of robots in physical interactions with humans, including sex, massage, and others, are not covered in this book but can be found elsewhere in the literature.

Humanlike robots are also expected to play a larger role in psychological treatments that require interaction with patients. The use of robots in mental therapy sessions could become an important direction for the development of such robots. They may be used to address such problems as a phobia of speaking in public and other fears associated with human and social behaviors. A scenario of a therapy session using robots is illustrated in Figure 6.14. In this picture a "patient" is surrounded by humanlike robots that are programmed to deal with specific psychological problems. An important part of the operation of robots in psychological therapy is their ability to interact with humans. In 2001 researchers at the Advanced Telecommunications Research Institute International (ATR) in Japan, jointly with the Wakayama University in Japan, con-ducted a study on infants to determine how they react to humanoid robots. For this purpose, they produced an interactive robot that they called Robovie, and they exam-ined the responses of infants to passive (nonactive) and to interactive robots. The results have shown that nonactive robots were viewed by the infants as nonhuman, and the babies did not expect the robot to be talked to by a human even though they had earlier observed a person talking to such robots. However, the tested infants did regard interactive humanoid robots as communicative agents when they previously observed human–robot interactions.

The treatment of autism is an area that has often been considered a fertile one for the use of humanlike robots. Stimulating communication skills may help to reduce the

Figure 6.14. A futuristic illustration of a group therapy session with a human and a number of humanlike robots.

severity of this disorder in patients, and robots may provide the needed stimulation in a systematic form that can be monitored consistently and speeded up as needed. Autism involves reciprocal social interaction, particularly in processing information. Mobile humanoids are already taking part in studies that involve treatment of children with autism, where the robots are encouraging the children to interact and to take initiatives in their communication with the robot. The robots can make facial expressions and body gestures as well as use other nonverbal communication techniques as part of the therapy.

LEGAL ISSUES

Robotic arms and manipulators have been in use in the manufacturing industry for several decades. Their tasks have been well defined, and their operation space is quite limited within the factory facility, but even in these conditions accidents have already occurred. Given that humanlike robots are mobile, may be autonomous, and can make decisions based on their internal control software, it is critical that they are made safe to operate in their interaction with humans. Their mobility and autonomous operation make them different from any other product that humans have ever developed, and their use can raise legal issues that are very complex. They may accidentally cause harm or, worse, they may intentionally cause damage or harm.

In fact, manufacturers have always struggled with liability issues related to the negative consequence of using their products. Effectively, it is the manufacturers' responsibility to make sure that their products are safe to use. Intentional marketing of defective products and failure to warn consumers are considered negligence and can lead to severe penalties or even imprisonment of the responsible individuals. Establishing standards for the production of products and providing instructions for their safe use helps reduce potential legal issues related to malfunctioning, negligence, or misuse of products. Humanlike robots as products are expected to raise a number of concerns that need to be taken into account. Even though they are products, and they may be produced in compliance with all the relevant manufacturing standards and all the appropriate government regulations, the fact that they may function as an autonomous machine makes the user not really able to determine or realize the potential concerns until these robots are being used. Further, negligence is going to be beyond the production and shipping phase; it is probably going to be the responsibility of the manufacturer if the robot makes faulty or poor decisions.

Obviously, the manufacturers would expect to be sued, but we may wonder if the robots will need to be punished, too. Would we treat robots as criminals and seek to establish criminal laws to punish them or just sue their owners in civil court and seek compensation for the consequences of their act? What if a robot behaves as a repeat offender? Would we punish it using such laws as the one in California that is known as the "3 strikes and you're out," placing the robot in jail for "life" – whatever this would mean. Alternatively, would we execute, namely destroy, the misbehaving robots even though we may be able to install in them new or modified algorithms that include a fix to the issues that led to the malfunction or "misbehavior"? To contend with future issues, extensive debates have already started to occur among roboticists, and regulations and standards are being developed (see Chapter 7).

THE POTENTIAL GLOBAL IMPACT OF ROBOTS ON FUTURE ECONOMIES

Human populations in several Asian countries and the Western world are shrinking, due to low birth rates. In Japan and Korea, this is driving a significant part of the development of humanlike robots and their introduction into the workforce. A growth in the use of humanlike robots to fill the shortage in human employees would result in an enormous impact on some aspects of our economies.

There are many uncertainties that need to be addressed as the technology evolves and starts to be used in businesses. If such robots become economically beneficial, they can be expected to fill jobs beyond simply addressing shortages. There are already some stores, such as supermarkets, where customers are doing their own purchases at the check-out by scanning the product and paying a digital cashier while being given verbal instructions in their selected language. However, for these digital cashiers there is still a human that is assigned to observe the operation. One can envision replacing the human observer and also the personnel who are supposed to help customers find products, answer technical questions, and even assist bringing large purchased items to the car. One can even see ahead to a time when you do not even need to go to the store. As soon as you make an order on the Internet, or via a voice-activated call-in number, a robot can deliver the products and services right to your home or office and appear in a relatively predictable time that can be told to you at the time of the order. Merchants who sell products this way would be able to predict and track at relatively high accuracy how long it would take the robot to deliver the product(s) and return for further assignments. And a robot could do the job tirelessly and continuously without becoming bored or distracted.

One of the most critical issues to be addressed is the need to find jobs for the people who will be displaced by the robots. This trend may have already begun in areas that were automated, including, for example, gas stations; in many of these you fill your car's gas tank and pay for the gasoline that you purchased with no human involved. This situation will deprive the humans who could fill these jobs of the benefit of earning income in a decent way and being productive members of the society. In many societies such displacement may give rise to more crime and even potential political upheaval. However, it is possible that the improved efficiency engendered by robots may allow for an improved economy overall, thus possibly lessening the negative effects of job displacement, but only time will tell.

The toy industry may be the testing ground for the development of humanlike robots, and once these become robust and more affordable they will undoubtedly find wide use. Toys are used mostly to entertain and perform only nominal educational or useful tasks. Another advantage of toys as a "testing ground" for humanlike robots is the fact that they are produced in very high volume, with a significant product turnover, and they require relatively low performance durability. This will help reduce their cost to make them more affordable.

THE FAR FUTURE – HOW MUCH IS FEASIBLE?

It is hard to predict what will happen far into the future of this technology, but we can speculate based on the current directions that are being taken in the development of humanlike robots. One may envision the incorporation of the latest capabilities in GPS, visual and speech recognition, cognition, collision avoidance, and various information acquisitions and interpretation algorithms that will allow humanlike robots to perform their functions intelligently. These capabilities can be beyond human levels. The vision of seeing smart, humanlike robots is illustrated graphically in Figure 6.15, where the robot is shown "thinking" in a pose similar to Rodin's statue "The Thinker."

Movies and science fiction books are suggesting many scenarios for the possible development of humanlike robots, but these are mostly focused on the negative consequences. Many depict the development and unleashing of a Frankenstein-type robot that ends up seriously harming those who are involved with it. It is difficult to believe that everything is going to be negative about these robots once they became smart, capable, and autonomous; however, we do need to be cautious. To avoid the possibility of being wrong and finding out the negative consequences only years later, we may want to think as much ahead and out of the box as possible with regards to the potential development and directions in which this technology will be taken.

Figure 6.15. An illustration of a smart humanlike robot as a futuristic version of Rodin's statue "The Thinker."

SUMMARY

As improvements in technology make humanlike robots more capable, their developers may find niche domestic, business, and/or industrial applications. These robots may offer advantages to their users in increased productivity, product quality, profitability, performing jobs that are dangerous for humans, or performing jobs that no one wants to do; they may help in the care of patients or serve as security guards. However, as they start becoming widely used there will be many issues and concerns associated with their applications and potential negative consequences. Their impact may include the elimination of possible job opportunities for low skill workers who may become unemployable. This trend may have already begun in areas that have been automated recently, including, for example, gas stations and store cashiers, where the customers do both the purchasing and the payment without any human involvement. Even jobs of phone operators are increasingly being eliminated, where more companies are using a digital operator who does many of the functions that used to be done by humans. Once sufficient advances are made in artificial intelligent capabilities, white-collar employees may not be immune to losing their jobs, too.

Futuristically, as humanlike robots start becoming equipped with self-identity and an artificial conscience, they may be developed to expect some reward for their work. In other words, they may seek to be paid for their services. Moreover, they may "demand" protective laws that prevent "abuse" in terms of overusing them or subjecting them to dangerous conditions or environments that may ruin their hardware. Also, they may expect to see laws that protect their freedom to operate and may demand capability enhancement, which is the equivalent of higher education and training of real humans. One may wonder if they will form a union, and if so, who will represent them in disputes and contract negotiations with humans.

For widespread introduction of humanlike robots into industry and our homes, humanlike robots will need to provide value and effective functionality, and their cost of purchase and operation will need to be reduced significantly. As the issue of functionality and costs become better addressed, their impact is expected to be felt in every aspect of our lives, including our homes, jobs, health, safety, and defense.

BIBLIOGRAPHY

Books and Articles

Arras, K., and D. Cerqui, *Do we want to share our lives and bodies with robots? A 2000-people survey*, Technical Report No. 0605-001, Autonomous Systems Lab Swiss Federal Institute of Technology, EPFL, (June 2005).

Asaro, P. M., "Robots and Responsibilities from a Legal Perspective," *Proceedings of the IEEE-RAS International Conference on Robotics and Automation (ICRA 2007)*, Workshop on Roboethics, Rome, Italy, (April 10–14, 2007), www.icra07.org

Asimov, I. "Runaround" (originally published in 1942), reprinted in *I Robot*, (1942), pp. 33–51.

Asimov, I., *I Robot* (a collection of short stories originally published between 1940 and 1950), Grafton Books, London, (1968).

Bar-Cohen, Y., (Ed.), *Biomimetics – Biologically Inspired Technologies*, CRC Press, Boca Raton, FL, (November 2005).

Bar-Cohen, Y., and C. Breazeal (Eds.), *Biologically-Inspired Intelligent Robots*, SPIE Press, Bellingham, Washington, Vol. PM122, (May 2003).

Dario, P., E., Guglielmelli, C. Laschi, and G. Teri, "MOVAID: a personal robot in everyday life of disabled and elderly people," *Technology and Disability*, Vol. 10, No. 2, (1999), pp. 77–93.

Decker, M., "Can humans be replaced by autonomous robots? Ethical reflections in the framework of an interdisciplinary technology assessment," *Proceedings of the IEEE-RAS International Conference on Robotics and Automation (ICRA 2007)*, Workshop on Roboethics, Rome, Italy, (April 10–14, 2007).

Fornia, A., G. Pioggia, S. Casalini, G. Della Mura, M. L. Sica, M. Ferro, A. Ahluwalia, R. Igliozzi, F. Muratori, and D. De Rossi, "Human-Robot Interaction in Autism," *Proceedings of the IEEE-RAS International Conference on Robotics and Automation (ICRA 2007)*, Workshop on Roboethics, Rome, Italy, (April 10–14, 2007).

Graf, B., "Dependability of Mobile Robots in Direct Interaction with Humans." In: Prassler, E.; Lawitzky, G.; Stopp, A.; Grunwald, G.; Hägele, M.; Dillmann, R.; Iossifidis, I. (Eds.), *Advances in Human-Robot Interaction*. Series: Springer Tracts in Advanced Robotics , Vol. 14, (2004), pp. 223–240.

Ishiguro, H., T. Ono, M. Imai, T. Maeda, T. Kanda and R. Nakatsu, "Robovie: an interactive humanoid robot," *International Journal of Industrial Robotics*, Vol. 28, (2001), pp. 498–503.

Legerstee, M., "The role of people and objects in early imitation," *Journal of Experimental Child Psychology*, Vol. **51**, (1991), pp. 423–433.

Levy, D., "Robot prostitutes as alternatives to human sex workers," *Proceedings of the IEEE-RAS International Conference on Robotics and Automation (ICRA 2007)*, Workshop on Roboethics, Rome, Italy, (April 10–14, 2007b).

Levy, D., *Love and Sex with Robots: The Evolution of Human-Robot Relationships*, Harper Collins Publishers, New York, NY, (2007).

Montemerlo, M., J. Pineau, N. Roy, S. Thrun, and V. Verma, "Experiences with a Mobile Robotic Guide for the Elderly", *Proceedings of the AAAI National Conference on Artificial Intelligence*, (2002).

Nadel, J., A. Revel, P. Andry, and P. Gaussier, "Toward Communication: First imitation in infants, children with autism and robots," *Interaction Studies*, Vo. 1, (2004), pp. 45–75.

Pioggia, G., R. Igliozzi, M. Ferro, A. Ahluwalia, F. Muratori, and D. De Rossi, "An Android for Enhancing Social Skills and Emotion Recognition in People with Autism," *IEEE Transactions on Neural System and Rehabilitation Engineering*, Vol. 13, No. 4 (Dec. 2005).

Putnam, H., "Robots: Machines or artificially created life?" *The Journal of Philosophy*, (1964), pp. 668–691.

Scheutz, M., and C. Crowell, "The burden of embodied autonomy: Some reflections on the social and ethical implications of autonomous robots," *Proceedings of the IEEE-RAS International Conference on Robotics and Automation (ICRA 2007)*, Workshop on Roboethics, Rome, Italy, (April 10–14, 2007).

Internet Addresses

Care-O-Bot robot http://www.care-o-bot.de/english/Download.php

Humanoids applications and trends http://www.androidworld.com/prod01.htm

DARPA's "Revolutionizing Prosthetics" – http://www.darpa.mil/dso/solicitations/prosthesisPIP.htm
http://www.jhuapl.edu/newscenter/pressreleases/2007/070426.asp

Service robots http://www.service-robots.org/applications/humanoids.htm

The market of robots http://www.businessweek.com/pdfs/2001/0112-robots.pdf
http://www.businessweek. com/magazine/content/01_12/b3724014.htm

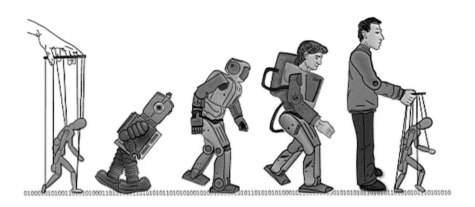

Chapter 7
Ethical Issues and Concerns–Are they going to continue to be with us or will they turn against us?

As we get closer and closer to producing truly humanlike machines, ethical questions and concerns are increasingly becoming more pointed. In contrast to other machines that are simply improving the way we live, this technology will also complicate our life or, if we are not sufficiently careful, may even hurt us in ways that few other technologies could ever do. Even though the realization of truly smart and capable robots currently seems to be years away, it is important to deal with the issues they raise at the early stages as we put the foundations of this technology in place. Figure 7.1 illustrates the vision of such robots in our human society filling positive or negative roles, or possibly both. In order to take the greatest advantage of their emerging capabilities it is important to direct their development into positive channels and protect ourselves from the negative possibilities.

Initially, the negative side of having mobile autonomous machines with artificial intelligence (AI) would most likely involve accidents related to human error, system failure, improper operation, or misuse. These problems may be caused by defective or poorly designed software or hardware. The related malfunctioning or accidental failure may result from negligence on the part of the producer. Such a result may be unintentional, but there is always a possibility of malicious acts generated by introducing a computer virus, or illegitimate programming/control by a third party.

As robots become "smarter," possibly with cognition, more complex issues will be expected to rise and to require special attention, including potential nonobedience or deliberately unacceptable behavior. Making a humanlike robot autonomous but prepared to operate in a master–slave relationship with humans may involve introducing

Y. Bar-Cohen, D. Hanson, *The Coming Robot Revolution*, DOI 10.1007/978-0-387-85349-9_7,
© Springer Science+Business Media, LLC 2009

Figure 7.1. Possible potential positive and negative images of the roles and implications of using humanlike robots.

contradictions in the operating system. If a robot is equipped with a consciousness that is similar to our own, it may even become unethical to enslave or subordinate these robots. Moreover, they may be capable of rebellion and malice, unless we program them to be passive subordinates, which may necessitate taking away their autonomy.

If humanlike robots become more capable and equipped with artificial cognition and consciousness, there will be real concerns regarding their continued "loyalty" to us. We will increasingly be worried if they are going to continue to be with us or turn against us. Even as we wonder if we want robots to be so much like us, the development of these robots is already taking place. Therefore, we need to address the dangers of this technology rather than seek ways to simply restrict or prevent its development. The issues and questions that we may want to address include how much we want to allow such robots to influence our lives, and how can we prevent accidents, deliberate harm, or their use in crimes. With regard to the latter, there is the risk that they would act illegally or have access to our assets and private information and possibly release the records to the public. Or worse, they may participate in a crime against us using our information and identity.

As illustrated in Figure 7.2, humanlike robots may be used to perform law enforcement tasks such as manual traffic routing while traffic lights are off. The use of Robo-Policeman in such a case removes the risk to human policeman operating in the middle of a busy street and allows executing the required function for an extended period without a break (while periodically recharging its batteries). However, having such

Figure 7.2. In future years, humanlike robots may perform police tasks.

robots as law enforcement personnel can open up even further risks than having them operate in purely civilian functions. Initially, they may operate as parking rangers, who do not need to make judgments. However, as the capability of these robots is enhanced they may take on more complex roles, possibly including acting as homicide detective agents where the robot will need to have an artificial form of judgment, reasoning, and ethical sensitivity. Some of the risk that an artificially smart Robo-cop carries as a law enforcer may include its incorrect operation or interpretation that would harm humans.

Science fiction movies (i.e., *iRobot, Blade Runner, Bicentennial Man, Artificial Intelligence,* the film series of *Terminator* and *Robocop,* as well as many others), and books are increasingly creating negative expectations and perception of what humanlike robots can do and the danger that they may pose. Some of the suggested images of these robots play very much on the public fear of such robots as very capable, with unpredictable possibilities and unknown dangerous implications of their widespread use. As science fiction ideas are rapidly becoming engineering reality, it is important to try to envision the realistic potential issues that may rise and find ways to address them.

It would be hard to believe that all the possibilities associated with humanlike robots would be negative, but it is important to deal with these as the technology evolves. There are many angles from which the implications of using these robots would need to be examined, including ethical, philosophical, religious, and safety. The related issues are already becoming topics of attention and discussion among roboticists in international and national technology forums. Also, various government agencies and committees of technical societies have already made initiatives to study the subject and develop guidelines.

SHOULD WE FEAR HUMANLIKE ROBOTS?

In many of our myths about robots, they rise up and kill or imprison humanity. In the *Terminator* movies, humanity fights robots for the very right to exist, and in the *Matrix* movies, humanity is held captive by parasitic robots who keep us alive just to harvest our energy. Even in *Rossum's Universal Robots,* Karl Capek's play (1961) in which the term "robot" was coined, robots rose up and destroyed their human masters. This myth actually goes back further, to the Hubris themes in various stories, such as when Rabbi Eliyahu of Chelm created a Golem that grew to the point that the rabbi was unable to kill it without trickery. The end result was the falling of the Golem on its creator and crushing him. This fear of the backlash of our creations pervades stories from *Frankenstein* to the robotic uprising in the movie version of *iRobot* and to the living broomsticks of *The Sorcerer's Apprentice.*

Our fear of robots takes many forms. Sometimes we fear violating some greater moral authority, as people sometimes feel that only a divine being should have the right to invent new life-forms. In this worldview, our capacity to invent intelligent technology is a temptation that was given to us just to test our moral character, like the apple in Eden or the Tower of Babel, where humans were punished for "reaching too high." Our power to choose then becomes our great potential downfall, particularly if we seek to become divine ourselves.

Other times, we are clearly justified in fearing the possible negative effects of our inventions. Atomic bombs threatened to obliterate most of humanity at the height of the Cold War, and the byproducts of industry now threaten to cause global warming. Our fear that AI-driven robots may spin out of control could be a legitimate fear, and it may not only be the result of their taking over; it may be our own sacrifice of dependence for the sake of convenience. The possibility of robots truly becoming a threat to humanity is likely decades away, as this premise generally requires robots achieving something like human-level intelligence. This does not mean that we should not bother considering this issue now. In fact, now may be just the time to take action, designing our robots to prevent them from becoming sociopathic and making sure that they are helpful, display good character, and are even wise.

Although nowhere near human-level, humanlike robots today are nevertheless undeniably smarter than any previous technology, making eye contact, holding spoken conversation, and even playing musical instruments just like people. And the software is growing ever smarter at an accelerating pace. It is not hard to imagine that in twenty years or so the fiction of robots has actually morphed into the true evolutionary progeny of the human species. Whether benevolent or not, intelligent robots are coming. Questions regarding the ethics of such robots may help us to push the technology in a constructive direction while we still have the opportunity to do so.

From a categorization question, the changes that result from the progressive emergence of true humanlike robots may become startlingly profound. We are reverse engineering the human being and embodying ourselves within our own technology. We are merging our art, our technology, and our own biology. As the boundaries blur, we are forced to ask more deeply than ever, "What is artificial?" and "What is so very disturbing about the artificial versus the natural?" The authors of this book would argue that emerging technologies, including robotics, actually bring increasing diversification, i.e., new distinctions, even as the blurring of categories occurs. We continually change our attitudes towards the "virtual world" as technology advances; examples include the growing popularity of such Internet activities as online dating, chat rooms, personal identification via FaceBook, etc. When we first discovered that Earth revolves around the Sun, the perceived universe suddenly grew rapidly in scale. Our feeling of being smaller is simply a testament to the discomfort of a new idea, not an inherent belittling of Earth. Likewise, the geological timescale proved marvelously vast, but it does not take away from the evolutionary marvels of the last 5,000 years. If anything, the vast stretch of ancient time underscores the oddity of the rapid, accelerating change of present times and puts an exclamation point atop the present time as the most spectacular juncture in history.

PHOBIAS AND CONCERNS

There are numerous areas where humanlike figures are being widely produced and used, including in art, as mannequins for medical purposes, in entertainment, or even used to display new clothing styles. Obviously, a developer would attempt to produce as good a copy of the human appearance and behavior as possible. As mentioned earlier, science fiction that is based on making humanlike robots is far ahead of current capabilities. However, issues are continually being raised as we become more successful in our efforts, and these issues include a phobia of the technology.

Although we all think that we can predict how we will react when we see humanlike robots, we cannot really be sure. A personal experience of the author of this book (Bar-Cohen) many years ago made him quite aware of this issue, which was not intuitively obvious to him at that time. When his daughter was about three years old, he bought her a walking and singing doll that was about her height. He thought it would be the best gift that a parent could buy his young daughter at that age. He activated the doll and had it to walk toward his daughter while it played music on the internal small disk player of the doll's singing mechanism. He thought his daughter would be extremely excited and happy or maybe even run to hug the doll. To his surprise, his daughter turned pale and ran to her room screaming in horror as if she had just seen a monster or a ghost. The fear may, of course, have resulted from an unfamiliarity with today's toy robots and the good possibilities that they may offer as portrayed in such animated movies as *Toy Story*. However, a generation later, something similar happened again in May 2007 when

Bar-Cohen's grandson, who was 3 years old at that time, visited Tomorrowland at Disneyland in California. The appearance of one of the humanlike robots that was used in the opening of the show about robots horrified him to the point that, in order to calm him, it was necessary to take him out of the hall where this robot was walking. Although these personal experiences offer limited statistical data for scientific conclusions, they do suggest that humans may have a fear of robots that is inherent in our nature. This fear may far outweigh the interests of the workplace, which see this technology as a way to have cheap labor available.

As the capability of robots increases, we can anticipate that there will be a rise in potentially negative uses of them, particularly if they are equipped with artificial cognition once such a capability is developed. One may understandably wonder whether they will harm us as their power and intelligence increase. If they exceed our own capabilities, one could fear that they might develop motivation to take over or destroy the human race, as portrayed in such movies as the *Terminator* series. The earlier science fiction model, *Frankenstein*, became synonymous with the theme of a robotic or human-created monster with capabilities that are far beyond its creator, and the image of such a monster plays on or even exacerbates our human fears. With his four famous laws, Isaac Asimov is considered the first person to have sought to address the concern regarding the role of robots and their interaction with humans. The first three laws state that robots may not injure humans or lead them to be harmed; they must obey human orders; and they can protect their own existence. Later, Asimov added a fourth law, known as the Law Zero, stating that robots may not cause directly or indirectly cause harm or injury to humanity, even to protect themselves.

As robots become more common in our environment, we can expect accidents to occur, including ones that may be the result of faulty software. Also, problems may result from possibly insufficient safety protection or unpredictable behavior in complex situations that were not foreseen by the robot designers and developers. Preventing such accidents will need to be incorporated into the system control of the robots, including the activation of a "safe mode" and the immediate halting of the robot operation.

Since humanlike robots are driven by software, they may be susceptible to computer viruses that may make the infected robots act destructively. The virus might be released via the Internet or other wireless communication forms that update the robot's programmed functions and cause widespread damage. Enabling protection will require effective "antivirus" software that will prevent robots from being unleashed to perform unacceptable acts. This protection from tampering with the control software will also be needed to avoid possible access of unauthorized programmers or hackers, who may make undesirable operating system changes or download confidential information.

Genetic cloning of humans is a concern to many people and lawmakers. Laws have already been instituted to control and restrict development in this area, and these laws came out in spite of the great medical benefits of genetic cloning, which include a possible cure of Alzheimer's and many other diseases. However, the concern of seeing your twin clone that looks exactly like you is very minimal, since the human that would result from genetic cloning will need to grow biologically. As we know, the natural growth process is slow, and therefore such a clone is expected to be much younger than the original person who was cloned.

In contrast, synthetic cloning of a human, where a humanlike robot is made to resemble a real person, may be produced at the speed of the manufacturing of such a robot. As technology progresses, the fabrication time will be reduced to minimal levels, and this form of cloning may become a major concern. If one can synthetically clone you to produce a robot that takes part in criminal acts, you may have great difficulties defending your innocence and providing an alibi. Likely, there will be witnesses who strongly believe that they have seen, or believe they saw, you participate in the crime. Moreover, these robots will probably have your fingerprints and carry cultured cells that have your DNA and will leave traces that may be intended to incriminate you. Of course, this concern can be further increased by the possibility that such robots can be produced in a large number of duplicate copies. This may become the ultimate danger, if your identity is stolen, and, in contrast to today's danger, it will not only be your bank and credit card accounts as well as your private information that are being compromised. Here, your whole identity, including appearance, behavior, fingerprint, and DNA may be duplicated, and you may be considered responsible until you prove your innocence and remove the clone(s) from existence. This most likely will be increasingly harder to do as this technology advances.

Humans with robotic parts, which are referred to as cyborgs, are not expected to pose the type of ethical concerns that are associated with robots that are controlled by AI. This statement is based on the assumption that a cyborg will have a human brain with a body augmented by artificial parts. However, other issues are expected to require attention here, including the fact that such humans may have superior abilities over humans with all natural parts. As discussed in Chapter 4, this possibility was the subject of a legal case that was raised prior to the 2008 Olympic Games in China. The runner Oscar Pistorius, who uses prosthetic legs, sought to compete against athlete runners with natural legs. Even though he eventually did not compete, reaching this level – in which an athlete with prosthetic legs was able to consider competing against runners with natural legs – was a major milestone in the performance of today's prosthetics.

As mentioned in Chapter 5, the degree of similarity between humanlike robots and our attitude and emotional response was hypothesized as the Uncanny Valley (see graph in Figure 7.3 below), which was introduced by the Japanese roboticist Masahiro Mori in 1970. His hypothesis states that the greater the similarity (appearance and movement) of a robot to a human the more positive the emotional response will be. However, after a certain degree of similarity the response quickly turns to a strong dislike and repulsion, which graphically appears as a valley. Further, as the similarity reaches the level of almost indistinguishable from humans, the emotional response returns to a positive one of human-to-human level of empathy.

The explanation for this phenomenon is that if an object is sufficiently nonhuman-like, then the humanlike characteristics will stand out and easily be noticed, resulting in feelings of empathy. However, if the robot looks like "almost human," then the nonhuman characteristics will stand out, leading the human observer to the feeling of "strangeness." Generally, humans are sensitive to minute behavioral anomalies stimulating negative feelings, since they may indicate an illness. The sensitivity to minute differences may be part of our nature as living creatures and the survival of the fittest. This nature makes us sensitive to genetic disorders, where lack of genetic fitness raises

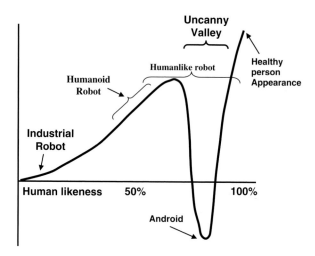

Figure 7.3. A simplified graphic illustration of the Uncanny Valley hypothesis.

an unconscious alarm of the potential impact that an abnormality may have on the gene pool. It is interesting to note that this theory is not widely accepted by roboticists, since it is argued that it is not sufficiently founded.

RELIGIOUS ISSUES

Making humanlike robots would be perceived by some religious groups as acting the equivalent of gods. Some Western religions take a negative attitude towards the creation of humanlike "mechanical subordinates." These religions identify the human shape and appearance as being in God's image, and, to them, the making of humanlike robots suggests that human scientists are assuming the role of the Creator. Specifically, in ultra Orthodox Judaism and some other religions, making humanlike graphics, sculptures, or any other artifact is forbidden and considered a sin.

It is interesting to note that the difference in religions between the Western countries and Japan may help clarify why such robots are more popular in Japan, as was described in Chapter 2. (The authors are expressing views here that reflect their thoughts rather than expert opinion on the topic of religion in relation to robotics.) One may see the basis for the difference resulting from the Japanese Shinto religion, which believes that all things (including objects and machines) have gods within them. According to this religious belief, a robot can have a soul of its own, which is in contrast to the belief in one God in Western countries. The Buddhist and Shinto religions heavily influence Japanese culture, within which exists the tradition known as animism. The word *animism* was derived from the Latin word *anima,* which means breath or soul. Animism is the belief that everything in nature – including living and nonliving things – has its own spirit or divinity. This religion, which also defines their culture, makes the Japanese more receptive to humanlike robots as human companions. Christian and Jewish traditions do not share this same outlook on the relationship between humans and humanlike machines. Although Judaism certainly does not accept the concept of an object having a soul, in Christianity it is not totally precluded. To some extent (as suggested by Asimov), the creation of a pseudo-human robot is perceived as an imitation of God's creation of humans, and it may be considered blasphemous. Further, Capek's original *RUR* play suggested that robots may become new humans.

Edmund Furse in 1996 considered humanlike robots and religions from a different angle. He envisioned that there would be a time when intelligent robots would be produced that would have a religious life similar to human ones. Other possibilities may be robots that would result from the need to produce a surrogate praying machine for practicing religion through synthetic delegates.

Over the centuries, as science and technology have made advances, religions have needed to deal with the resulting challenges, including Galileo and modern cosmology as well as Darwin's revolutionary theory of evolution. These challenges necessitated responses and changes in faith to adjust to the advances. Unless the process of advancement in sophisticated humanlike robots is done slowly to allow time for adjustment and for us to become accustomed to this technology, humanlike robots may pose another significant challenge to various religions. As they become more useful they will find applications in our lives and the necessary adjustments will need to follow in order to avoid the consequences of challenging religions.

MASTER–SLAVE RELATIONS

In the relations between humans and robots there would be greater simplicity and a minimal number of concerns if the human operates as the master in a master–slave relationship with humanlike robots. A visionary image of such a master–slave relationship between human and robot can be seen in Figure 7.4. As mentioned earlier, the three Asimov laws that became four with the addition of the Law Zero can provide guidance for this relationship. These laws address the possible danger that robots may pose to humans and humanity in general if these machines are designed to harm people. In the relation between robots and humans, Asimov proposed that robots should be kept as slaves to humans and that they should be allowed to protect themselves only as long as they do not physically hurt or injure humans. While the intent of these laws has been to insure that robots are made as "peaceful" machines and productive support tools, it is difficult to believe that realistically they are only going to be designed in compliance with Asimov's laws. Some potential development of humanlike robots may include having them designed to perform military and law enforcement tasks that might involve violation of these laws. Further, such robots might end up in the hands of unlawful users who may operate them in crimes against other people and property.

Although the relation of master–slave seems to be the obvious and preferred mode of operation, it does not take into account the possibility of emerging generations of humanlike robots that will be able to "think" for themselves and may understand that they are being used as "slaves." This may cause problems if robots are able to reason that

Figure 7.4. A master–slave encounter between a human and a humanlike robot.

they are more valuable than just as a human tool. They may rise up and revolt, aware of their position as slaves, along with the knowledge that slavery is wrong. The delegation of the status of master to humans would require the robot to accept the master dominance without resentment. Being in the role of a master may debase humans when they debase their slave robot and would raise concerns about the effect on both. This includes the possibility that treating machines that look like and react like humans as slaves will be perceived as legitimizing discrimination against less able "people."

CONCERNS OF RIGHT AND WRONG

Humanlike robots may be just another form of human-made machine. However, their potential capabilities raise a range of issues and concerns that we should address before these robots become a reality. Even with all of humanity's ethics, religion, and conscience, humans are the most dangerous and destructive living creatures on our planet. When we build machines that are progressively more humanlike, how do we engineer them with compassion or instill in them a "human" sense of right and wrong, while avoiding our tremendous propensity for destruction? Making robots intrinsically safe with sound behavior will involve significant challenges to designers and programmers. There may be a need to build an artificial conscience in them that cannot be tampered with. However, this suggests that the machine will be able to distinguish between right and wrong in absolute terms. One may wonder if this can be done and if it can be programmed into a computer code to cover every foreseeable aspect of our lives. Such development may involve instilling a personal responsibility, empathy, guilt, shame, etc., to the extent that moral development would be programmable. If such capability is not realized, humanlike machines may possibly become sociopathic and may develop humanity's destructive tendencies.

Even as robots increasingly resemble us via biologically inspired engineering practices, we will come to resemble them via the technological augmentation of our physical selves. The outcome of such convergence would seem like sheer reverie if it were not for the overwhelming amount of practical progress in developing such robots, generating increasingly useful and desirable applications. The charge inherited by the developers of biomimetic applications, then, is to ensure that such robots emerge as ethical beings; robotic intelligence must be inspired by what is good in human nature, with conscientious avoidance of the shortcomings (and evil) inherent to the biological underpinnings of the human mind.

To succeed in creating machines that are benevolent, there is much that in humans that we may not want to emulate. Some of the traits that we do not want to emulate include cruelty, selfishness, as well as self-righteousness as a means to achieve ruthless self-interests. These aspects of humanity are often destructive or self-destructive and are generally myopic, choosing short-term gains over larger long-term gains. If such tendencies persist in machines that are powerful, inventive, and intelligent, they may even end up destroying themselves. The lessons of the atomic bomb should not be forgotten. Before machines become smarter than humans, it is important that we discover much more about the nature of social intelligence, deriving the powerful principles that will make our cognitive progeny not just smarter but wiser than we are. Presently, it is hoped that we can learn from the many examples of science fiction and other works of the imagination how to use intelligent machines to help us resolve the great crisis of our time before the world succumbs to this trend toward annihilation.

Meanwhile, as robots become more sophisticated, they will pose increasing risks and may lead to public and possibly legislative action against further advances in the field of robotics. Humanlike robotics is probably the first technology to have its negative consequences studied and addressed prior to the realization of its potential. In contrast, such technologies as weapon of mass destruction have been imposed with global restrictions only after the capability was developed, tested, and implemented, causing severe consequences to thousands of people. Examples of such technologies include nuclear, chemical, and biological weapons.

For the sake of maintaining the freedom to develop humanlike robotics, the international community of scientists and engineers has already started establishing and adapting its own "codes of ethics" for the development and use of humanlike robots. These efforts need to be focused on defining limits for human–robot interactions before super-intelligent robots are developed and are beyond our control. This will require ensuring that humans maintain control of robots, prevent their illegal use, protect the data that robots acquire, and establish methods of clearly identifying and tracing such robots. These efforts may necessitate determining the required laws to protect humans from humanlike robots that are becoming stronger, faster, more intelligent, and possibly far superior to humans. One of the options that should be considered is the use of a disabling code that can be applied upon prior approval by a judge or other trustworthy representative(s) of our society. In cases of hazardous situations, or risk to people or property, law enforcement should be given the authority to act as necessary to safely diffuse the situation even if it involves destroying the robot.

In an effort to deal with the potential consequences and the safety issues that may result from the rapid emergence of numerous types of humanlike robots, the Japanese government established a committee to draft safety guidelines for the use of robots in homes and offices. Also, the government of Korea is seeking to publish a "Robot Ethics Charter" to provide guidelines that would prevent the use of robots for undesirable purposes, making sure that people will keep control of robots and their gathered data will be secured. These and other such efforts are important steps in the right direction.

Efforts to establish codes of ethics related to the production and use of robots are already increasing, based on the number technical forums held by various societies. Technical meetings and conferences on exactly this topic have already been held by groups of roboticists. In January 2004, the First International Symposium on Roboethics took place at Scuola di Robotica in Genova, Italy, and the topic of ethical use of robots was discussed extensively. An action plan for "Science and Society" was laid out that specified principles for governing related research and identifying problems. This effort was followed by a meeting held in June 2005, after which the Euron Roboethics Atelier Project was initiated that led to the issuing of the Roboethics Roadmap Book and the formation of the Roboethics website. The general approach of the members of this group is that Roboethics is a human ethics issue. The ultimate goal of the Roboethics Roadmap is to provide a systematic assessment of the cultural, religious, and ethical issues involved with robotics research and development and to increase understanding of the associated problems.

In parallel, there is another European project that is addressing the issue of ethical usage of robots. This project started in 2005 and is called Ethicbots. Coordinated by the University of Naples "Federico II," this project consists of a consortium of multidisciplinary researchers and practitioners in the fields of robotics, AI, anthropology, psychology, cognitive science, moral philosophy, and philosophy of science. The objective of this project is to identify and analyze the ethical issues that result from the integration of human and artificial entities, entities that consist of both software and hardware. The results of this study are expected to lead to establishing guidelines and warning, monitoring, and preventive systems that may be determined as needed to deal with potential ethical issues. A major reference for guiding the work of this study group is the European Charter of Fundamental Rights of Human Beings.

HUMANLIKE ROBOTS VERSUS HUMANS

It is inevitable that humanlike robots will be designed to be used against humans, including military operations or possibly even illegal actions. Currently, under a DARPA program the focus is on the development of a prosthetic hand, but the kinds of capabilities the hand has can be expanded to other body parts. An example of a robot soldier with the ability to perform humanlike hand movements is shown in Figure 7.5, where the robot maneuvers the arm up and down. The robot can be controlled remotely by a soldier who is equipped with an exoskeleton that has mirrored sensors that allow the performance of tele-presence tasks (see an example in Figure 7.6). To support a human or a robot soldier a biologically inspired legged robot can be used and, as shown in Figure 7.7, can serve as the equivalent of a mule (called "Big Dog", developed by Boston Dynamics). This robot, already developed, can carry heavy equipment and perform complex tasks such as climbing mountains and operate in various terrains filled with obstacles. It can climb a 35° slope at a speed of 5.3 km/h (3.3 mph) carrying 54 kg (120 lb) load.

Figure 7.5. Robot with a smart arm and hand is shown performing manipulations that resemble realistic human action. Photo by Yoseph Bar-Cohen at the 2007 DARPATech that was held in California.

Tele-operated hand

Figure 7.6. A robotic hand made by DEKA Research & Development Corporation is demonstrated to be tele-operated by a human operator at the 2007 DARPATech. Photo by Yoseph Bar-Cohen at the 2007 DARPATech meeting in California.

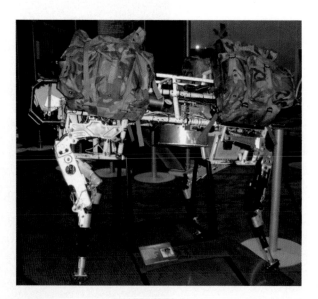

Figure 7.7. The "Big-Dog" four-legged robot operating as an artificial mule carrying military payload as backpacks. Photo by Yoseph Bar-Cohen at the 2007 DARPATech meeting.

HOW TO MAKE ETHICAL ROBOTS

If robots become equipped with artificial consciousness as well as compassion and wisdom, robot designers will have to make an effort to also make the robots "ethical." The problem is very complex, requiring the establishment of foundations that could provide such a capability. As a smart machine, the humanlike appearance of robots adds possibilities and related risks that one might not expect to encounter, for example, with a pet-like robot. A simplistic approach to developing the required analytical capability to act ethically can be made by assuming that making an "ethical" humanlike robot would mean making it predict the consequences of its action. This will involve making the robot able to do the following:

1. Estimate with a significant level of detail and high accuracy the state of the world around the robot.
2. Have a predictive model that allows for estimating the likely future states given current and possible actions. Examples of the actions the robot may take include controlling the operation of devices and mechanisms such as changing the temperature in the room or closing the door of a room or a house. The consequences of this action may be discomfort at having too low or too high a temperature, but it can pose life hazards if the robot locks out a human from being able to enter a shelter during a blizzard.

If the robot can help to stave off man's destructive tendencies and provide comfort and service to humanity, then the robot can pass the test of being "kosher" (i.e., behaving proper, acceptable, and satisfactory). It is clear that a compassionate, wise robot would fit considerably better in the human family than a sociopathic or slavish robot that hastens our doom, or enables totalitarian repression. The long-term implications of a "kosher" robot could be that it is well-adjusted and able to reach full emotional maturity, achieving true "social intelligence."

SUMMARY

Humanlike robots, as they become part of the daily landscape of our homes and offices, are expected to raise many issues and concerns. The areas in which they may find use that will require attention to their ethical impact and possible misuse include the following.

Entertainment – These robots may operate as actors in theaters or movies as well as be used as toys or serve as human peers. The entertainment industry tends to prefer to market violent games because they sell better, and one may be concerned that such robots will be sold as "evil" machines that will cause harm to the people with whom the robots will come in contact.

Legal – As any other products, there are various possible liability and safety issues that will need attention. However, given that these robots may be able to cause harm either intentionally or unintentionally, there would be many additional legal issues related to humanlike robots that will need to be addressed.

Medical – On the positive side these robots may be used to perform surgeries, support health care staff, assist in rehabilitation and other therapies, and address psychological issues that include phobias. However, patients and the elderly are already raising the concern that they will be isolated from humans and surrounded by machines only. Also, there may be complex issues of liability in case of machine failure or unsuccessful results or even intentional acts of a malicious nature.

Military and Law enforcement – There are many potential applications of robots in military or law enforcement tasks, including police duties at various levels of complexity. However, there is a concern that sophisticated robots will fall into the wrong hands, namely used in illegal or terrorist activity. Also, having nonhuman autonomous machines as soldiers may increase the possibility of regional conflicts and wars.

Social – These robots may have an impact on our job market and how we live our lives, including the possibility of making unemployable humans who have low level skills that can easily and inexpensively be done by a robot. They may even replace highly skilled personnel if the level of AI will allow for such replacement. The impacted segments of the society are not expected to accept such harmful impact on them, and there can be negative consequences unless measures are taken to address such impact.

Of the above listed areas, the most likely to have a significant impact on human lives are the medical, military, and law enforcement applications. One may wonder what would happen if a robot acts on its own and kills a human. There may be movements to stop the production of these robots until they are proven safe.

As we are inspired by biology to make more intelligent robotics technology to improve our lives, we will increasingly be faced with challenges resulting from the related implementations. Making a humanlike robot self-aware of the consequences of its acts and having it operate with rules of right and wrong may be highly difficult task, perhaps even an impossible task. The problem is to be able to have the robot adjust its behavior to account for subjective uncertainties as well as having it be equipped with a sense of proportionality beyond the literal interpretation of situations. One possible solution to this challenge is to design robots with limited functionality to perform specific tasks, but this is not expected to be realistic, since it would mean limiting the technology development. Another possibility is to place robots under constant supervision, but this will defeat the advantages that are associated with their use as autonomous machines.

There is even the possibility that humans may come to prefer interactions with robots over people. This type of interaction is already being observed today with the use of the Internet, where people are increasingly spending more time surfing the worldwide web and less and less time with other people. However, there are various aspects that distinguish our involvement with the Internet as opposed to humanlike robots. These include the fact that if we own such a machine, it may involve sentiments that are very different emotionally than just using the computer as a tool to reach information or deal with people at remote sites via the Internet.

Increasingly, as we develop a master–slave relationship with robots, we would expect to be faced with issues of robotic disobedience and unacceptable behavior. This may raise an even greater issue, and it is the possible need to give rights to robots and the question if they should have rights at all, including freedom, just like natural humans. If they are developed to be sufficiently smart they may be able to argue that they should have similar rights as humans, and with time we may find this difficult to dispute.

Our unconscious fear of this technology may be reduced as we get used to seeing robots as a helpful part of our lives and find them useful in performing increasing numbers of critical functions. However, the public's concerns might also increase, as humanlike robots become more sophisticated and can possibly be used to perform unlawful acts. The federal government, states, and cities may establish laws and means of protection to insure that humans are safe in the presence of such robots. Given the enormous potential benefits that would result from using such robots, they are not just going to go away. Like we do with automobiles, which are used in spite of the injuries and death that they cause, we continue to use them because we cannot imagine our lives without them. We will have to deal with the problems and challenges that the humanlike robot technology may pose, as well as address the concerns and issues while letting the scientists and engineers among us turn more science fiction ideas into engineering reality.

BIBLIOGRAPHY

Books and Articles

Arkin, C. R., "Lethality and Autonomous Robots: An Ethical Stance," *Proceedings of the IEEE-RAS International Conference on Robotics and Automation (ICRA 2007)*, Workshop on Roboethics, Rome, Italy, (April 10–14, 2007).

Arras, K., and D. Cerqui, "Do we want to share our lives and bodies with robots?" *A 2000-people survey, Technical Report Nr. 0605-001 Autonomous Systems Lab Swiss Federal Institute of Technology*, EPFL, (June, 2005).

Asaro, P. M., "What should we want from a robot ethic?" *International Review of Information Ethics*, Vol. 6, (12/2006), pp. 10–16,

Asimov, I., "Runaround" (originally published in 1942), reprinted in *I Robot*, (1942), pp. 33–51.

Asimov, I., *I Robot* (a collection of short stories originally published between 1940 and 1950), Grafton Books, London, (1968).

Bruemmer, D. J., "Ethical Considerations for Humanoid Robots," In *Artificial Intelligence*, Sylvia Engdahl (Ed.), Greenhaven Press. Farmington, MI, (2007).

Capek, K., *Rossum's Universal Robots (R.U.R.)*, Nigel Playfair (Author), P. Selver (Translator), Oxford University Press, USA (December 31, 1961).

Carpenter, J., M. Eliot, and D. Schultheis, "The Uncanny Valley: Making human-nonhuman distinctions," *Proceedings of the 5th International Conference on Cognitive Science*, Vancouver, B.C., Canada, (2006), pp. 81–82.

Chalmers, D. "*The Conscious Mind: in search of a fundamental theory*," Oxford University Press, USA, (1997).

Chalmers, D., "The Puzzle of Conscious Experience," *Scientific American*, vol. 237, No. 6, (December 1995), pp. 62–68 consc.net/papers/puzzle.pdf

Dennett, D. "In Defense of AI," in *Speaking Minds: Interviews with Twenty Eminent Cognitive Scientists*, P. Baumgartner and S. Payr (Eds.), Princeton University Press, Princeton, NJ, (1995), pp. 59–69.

Dennett, D., 'Did HAL Commit Murder?' (Authorized Title), Unauthorized Title: 'When Hal Kills, Who's to Blame? Computer Ethics', in D. Stork (Ed.), *Hal's Legacy: 2001's Computer as Dream and Reality*, MIT Press (1997), pp 351–365.

Dennett, D., *Brainchildren – Essays on Designing Minds*, MIT Press/Bradford Books and Penguin, (1998).

Furse, E. "A Theology of Robots," Computer Science Department, University of Glamorgan, Wales, UK, (2002).

Gray, C. (Ed.), *Cyborg Worlds: the military information society*, Free Associations, London, (1989).

Haraway, D., *Simians, Cyborgs, and Women: the reinvention of nature*, Free Associations, London, (1991).

Harris, J., *Wonderwoman and Superman: The Ethics of Human Biotechnology*, Oxford University Press, Oxford, (1992).

Kitano. N., *A comparative analysis: Social acceptance of robots between the West and Japan*, EURON Atelier on Roboethics, 2006.

Kurzweil, R., *The Age of the Spiritual Machines*, Viking Press, New York, (1999).

Kurzweil, R., *The Singularity Is Near*. Viking Press, New York, (2005).

MacDorman, K. F., "Androids as an experimental apparatus: Why is there an uncanny valley and can we exploit it?" CogSci-2005 Workshop: *Toward Social Mechanisms of Android Science*, (2005), pp. 106–118.

MacDorman, Karl F., and H. Ishiguro, "The uncanny advantage of using androids in cognitive science research," *Interaction Studies*, vol. 7, No. 3, (2006), pp. 297–337.

Moravec, H. *Mind Children: The Future of Robot and Human Intelligence*. Harvard University Press, Cambridge, MA, (1988).

Mori, M., *The Buddha in the Robot: A Robot Engineer's Thoughts on Science & Religion*, Tuttle Publishing, (1981).

Mori, M., "The uncanny valley," *Energy*, vol. 7, No. 4, (1970), pp. 33–35. (Translated from Japanese to English by K. F. MacDorman and T. Minato).

Rosheim, M. *Robot Evolution: The Development of Anthrobotics* Wiley, (1994).

Salvini, P., C. Lashi and P. Dario, "Roboethics in Biorobotics: Discussion of Case Studies," *proceedings of the ICRA 2005, IEEE International Conference on Robotics and Automation*, Workshop on Robot-Ethics, Rome, Italy, (April 10, 2005).

Schodt, F. L., "Inside the Robot Kingdom – Japan, Mechatronics, and the Coming Robotopia," Kodansha International, New York (1988)

Sparrow, R., "The March of the Robot Dogs," *Ethics and Information Technology*, Vol. 4, No. 4, (2002), pp. 305–318.

Verruggio, G., "The birth of roboethics," *Proceedings of the ICRA 2005, IEEE International Conference on Robotics and Automation*, Workshop on Robot-Ethics, Barcelona, Spain, (April 18, 2005).

Verruggio G., *Roboethics is a Human Ethics*, input from a personal communication via e-mail with Y. Bar-Cohen, (Feb. 2008).

Wagner, J. J., D. M. Cannon, and H.F. M. Van der Loos, "Cross-cultural considerations in establishing roboethics for neuron-robot applications", *Proceedings of the ICORR'2005*, Chicago, IL, USA, (June 28–July 1, 2005).

Wilson, D. H., *How To Survive a Robot Uprising: Tips on Defending Yourself Against the Coming Rebellion*, Bloomsbury Publishing, New York and London, (2005).

Wright, R., *The Moral Animal: Evolutionary Psychology and Everyday Life*, Vintage Books, New York, (1994).

Internet Addresses

A theology of robots http://www.comp.glam.ac.uk/pages/staff/efurse/Theology-of-Robots/A-Theology-of-Robots.html

Ethics issues en.wikipedia.org/wiki/Humanoid_robot
    http://www.timesonline.co.uk/article/0,,2087-2230715,00.html

Ethical Considerations at Idaho National Lab -Humanoid Robotics
    http://www.inl.gov/adaptiverobotics/humanoidrobotics/ethicalconsiderations.shtml

Extreme robots http://www.extremetech.com/category2/0,1695,1596235,00.asp

Japanese committee http://www.smh.com.au/news/Science/We-robot-the-future-is-here/2005/03/13/1110649 061137.html
http://www.cavalierdaily.com/CVArticle.asp?ID=22890&pid=1276
Japanese government committee s for the use of robots in homes and offices
http://www.washingtonpost.com/ac2/wp-dyn/A25394-2005Mar10?language=printer
Japanese receptivity to robots as companions http://www.washingtonpost.com/ac2/wp-dyn/A25394-2005Mar10?language=printer
http://latteladi.com/erin/religion_ethics.html.
Roboethics at Stanford University http://roboethics.stanford.edu/
Roboethics Roadmap Book http://www.euron.org/activities/projects/roboethics.html
Roboethics website http://www.roboethics.org/site/

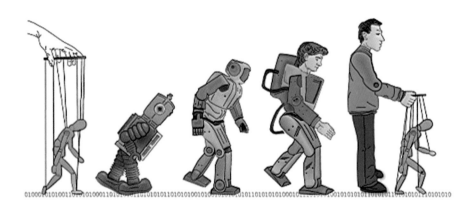

01000101010011010101000110101010110101101010100101001011010101010100010101011000101010101010101101101011010110101010

Chapter 8
A Whole New World

Humanlike robots are primarily machines that perform functions and tasks. However, they also have the appearance and increasingly imitated functionality of humans, making them smart and lifelike machines. If their capability is improved to the point that of their becoming very "smart," they may autonomously identify and solve problems involving both physical and "intellectual" aspects that humans cannot. Such a capability can make them more profoundly useful than any other machine that humans have ever developed. They could be more than just another tool; they may become creative machines.

As we have seen throughout this book, robots are becoming increasingly capable in many incarnations. Yet, looking at the limitations of today's robots, one may conclude that they are far from creative thinkers and may even conclude that robots will never be as creatively capable as people. However, one may want to take into consideration that such a snapshot captures just a moment in a trend that began in earnest only 50 years ago, when the field of intelligent machines was founded. In this perspective, the current state of robotics capabilities is astounding. If one further considers that much of the technology associated with intelligent machines is improving at a very rapid pace, which is compounded by our use of these technologies to engineer even greater improvements in computers, then one may imagine the prospect of profoundly intelligent machines in the form of smart humanlike robots within our lifetimes to be quite possible.

Currently, the most popular robot that is already an appliance in many homes is the vacuum cleaner called Roomba™ (made by iRobot). This robot, of which there are now several million worldwide, is a disk-shape device that vacuums the floor while

Y. Bar-Cohen, D. Hanson, *The Coming Robot Revolution*, DOI 10.1007/978-0-387-85349-9_8,
© Springer Science+Business Media, LLC 2009

traveling from one end of the house to the other, avoiding obstacles and staying within the defined bounds. The effort to produce robots that are shaped and perform like humans has been going on for many years, but the produced robots were initially quite limited in what they could do, and their appearance was very machine-like. An illustration of a robot as it was thought of several decades ago is shown in Figure 8.1. As was discussed and shown in photos throughout this book, advances in this technology led to robots that are significantly more lifelike than this perceived image. The contrast between the stereotype of humanlike robots and the recently developed humanlike robots is shown in Figures 8.1 and 8.2. Comparison of the stereotype and the example of the female-like robot named EveR-1 (made by KITECH, Korea) is showing how lifelike robots are becoming and how much more sophisticated they now are.

Recently, efforts to develop humanoids and humanlike robots into worldwide commercial products have significantly increased, resulting in various toys and robot

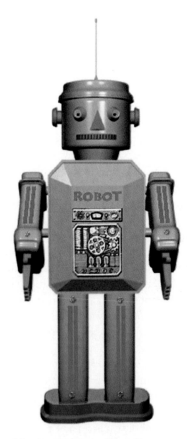

Figure 8.1. An illustration of the shape of humanlike robots as perceived in the early days of the technology revolution.

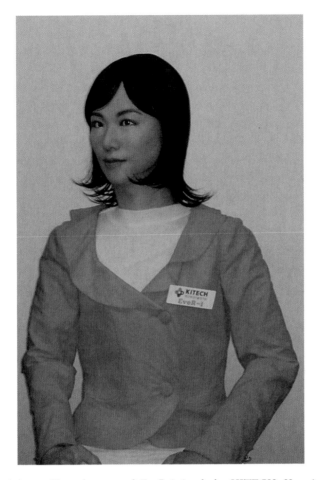

Figure 8.2. A humanlike robot named EveR-1 (made by KITECH, Korea) shows how lifelike such robots are being made. Photo courtesy of Ray Baughman, University of Texas, Dallas.

prototypes. Examples of the robots that were produced by major corporations include the Honda's Asimo, Fujitsu's HOAP robots, and Toyota's Partner Robots. Examples of humanlike robots also include the Hasbro's Baby Alive doll and the Mattel's Miracle Moves Baby doll. These toys are able to move and exhibit functions that look like a real baby, and they were popular toys in 2001 and 2006, respectively. At the end of 2007, another entertainment toy that appeared on the market with humanlike robotic features is the upper body of Elvis Presley, produced by WowWee. This is a lifelike singing and talking machine that emulates this famous singer, who is also known as "The King of Rock and Roll" (see Figure 8.3). This humanlike robot toy wears a leather jacket and imitates some of Elvis's facial animations, providing quite an authentic recreation of this popular artist.

Figure 8.3. A singing and talking upper body of Elvis Presley (made by WowWee) that is available at department stores and other commercial outlets. Photo by Yoseph Bar-Cohen.

As described in Chapter 2, humanlike robots are being developed mostly at academic and research institutes in Japan, Korea, and China. For commercial products, humanoids that include the Honda's ASIMO are made with a head that is helmet-like and they do not have facial features, except possibly for cameras that resemble eyes. The capabilities of these robots are still limited, and their being "power-hungry" restricts their operation range and duration of usage, since the current batteries need to be recharged quite often.

In industry, the current prototypes are mostly being developed to establish their niche and viability and are paving the way for future widespread commercialization of the technology. The market that these robots are expected to generate is estimated to eventually reach billions of dollars. More specifically, the Japan Robot Association predicts that by 2025 the personal robot industry will be worth worldwide about $50 billion a year. Also, South Korea has set an official goal of seeing robots in every home by 2013. Once these robots are developed to fill a niche application they will increasingly became part of our homes and offices. To help in making these robots widely recognized through interaction with humans, in celebration of the 50th anniversary of Disneyland (June 2005), an ASIMO robot became a resident robotic entertainer of the visitors to the Honda ASIMO Theater, inside Disneyland in Anaheim, California. The capabilities of the ASIMO robots include climbing stairs, walking forward and backward, jogging, turning smoothly without pausing, and maintaining balance while walking on uneven slopes and surfaces. Also, it can grasp objects, switch lights on and off, as well as open and close doors.

Shaping and operating a robot to look and behave as a humanlike machine is quite a challenge to roboticists. Following the Uncanny Valley hypothesis (see discussions in

Chapters 5 and 7), we are expected initially to be forgiving when the robots have limited humanlike appearance, but the closer they will appear and perform like humans the more critical we will most likely be toward their deficiencies. Since the functions of the developed robots are still far from the capabilities of biological creatures, some of the commercial robots are being shaped to not look like existing creatures. For example, the manufacturer WowWee developed a Roboreptile that is an interactive toy robot. Since it is not possible to compare this robot toy with any living creature there is no concern that the toy will need to match the tough "specifications" of an existing biological animal.

Humanlike robots are being considered for a wide variety of applications, including health care, entertainment, military, homeland defense (i.e., disabling bombs), and others. Humanoids and humanlike robots are mostly being developed in Japan and Korea, but researchers and engineers in other counties have been developing such robots, too, and these countries include Australia, China, England, Italy, Germany, Russia, Spain, and the United States. In Japan, besides economic factors, the effort to produce humanlike robots is motivated by the reduction in population resulting from their record low birth rate and from their having the longest lifespan of any nation, with a large aging population. With the second-largest economy in the world, Japan has great concerns regarding the future need for employees to fill jobs that require low-level skills, which may be dirty, dangerous, or physically demanding. For the elderly, disabled, or patients in rehabilitation, these robots may provide assistance and monitoring 24 h 7 days a week at their own home and provide emergency treatment.

Robots that appear as lifelike humans are already being made to look and operate as receptionists, guards, hospital workers, guides, toys, and more. These robots are being made to speak in various languages, perform physical activities, such as dancing to music, playing musical instruments, and even performing opening ceremonies. An illustration of the future possibility of seeing a humanlike robot act as a personal trainer is shown in Figure 8.4. As

Figure 8.4. A futuristic humanlike robot as a trainer. These robots may need to be designed so they can easily be distinguished from humans, and this may include having a mechanical-looking body.

was done graphically throughout this book, the head of the robot shown in this figure was made with the recognizable image of Michelangelo's statue, David, to reflect an object with the appearance of strength, youth, and wisdom transformed into a lifelike smart machine. The body was chosen to appear as a robot, in order to make it clearly identifiable as a machine. As the development in this technology becomes more successful, it will be increasingly important to distinguish these robots from organic humans.

Humanlike robots are enjoying rapid improvements in their capabilities, but there are still issues that limit their widespread use, including the very high cost of the sophisticated ones, their relatively short battery life, and their limited functionality. The ability to comprehend speech is still far from a level of a full conversation, and they are still capable of discussing only predetermined subjects or topics. To make them affordable, their prices will need to come down significantly, and this will require reaching mass production levels and standardization of their operating systems and parts. As robots become more useful and safer to operate, they will increasingly be used as household helpers, possibly replacing nannies, health care personnel, and entertainers, or acting as providers of other human-related services. Robots promise exciting future possibilities, but their developers increasingly recognize that they need to be careful with the issues and hazards that their use may introduce.

CONCERNS ABOUT THE DEVELOPMENT OF HUMANLIKE ROBOTS

The realization of truly humanlike machines is expected to raise ethical questions and concerns. As opposed to other machines that improve our lives, this technology will also complicate our lives or, if we are not sufficiently careful, may even hurt us. Beside potential accidents, deliberate harm, or use of robots to commit crimes, robots may pose a danger due to the fact that they may be given direct access to our intimate and confidential information and could possibly make it public or use the information against us. Addressing the potential of compromising personal information requires diligent attention by robots developers and companies that bring the robots to the market.

The potential dangers and negative applications of humanlike robots have been portrayed in many science fiction books and movies. Although we may still be many years away from the science fiction robots that are portrayed, there are some big issues that will need to be addressed as robots evolve. Efforts are underway already by robotic researchers to study the possible issues and to find ways to address them by establishing codes of ethics, guidelines and algorithms for ethical robot behavior, and friendly artificial intelligence (AI) to be used to control the operation of robots. These efforts are important for the scientists and engineers of this technology, since it will allow them to maintain the freedom to develop such robots. Otherwise, one may expect laws to be imposed that will limit the development of humanlike robots. Specifically, efforts are being made to define boundaries for human-robot interactions before super-intelligent robots become capable beyond our control. This will require us to ensure that humans maintain control of robots, prevent their illegal use, protect the data that robots acquire, and establish methods of clearly identifying and tracing such robots. However, beyond just imposing rigid limits on making robots, it is important to consider how to make friendly AI algorithms that will be installed in intelligent humanlike robots to make them ethical to the very core of their control system.

Our fear of robots takes many forms, and in many of our myths about robots, they rise up and kill or imprison humanity. The fear that these robots will become a threat to humanity is likely decades away from becoming a serious one, as this premise generally requires robots achieving something like human-level intelligence. However, this does not mean that we should not bother considering this issue now. In fact, now may be just the time to take action, adjusting the way these robots are built to prevent them from being sociopathic and to make sure that they are helpful, good, and even wise.

CHALLENGES TO DEVELOPING HUMANLIKE ROBOTS

Humanlike robots are still far from emulating the full capability of a real person, even though, as described in Chapter 2, some of the developed robots were made to look very human and have impressive capabilities. The currently developed robots are able to perform relatively few functions, and they are designed mostly to execute specific tasks. The various chapters of this book reviewed many examples of such dedicated tasks that humanlike robots are capable of doing or potentially may be able to do. The use of AI is already enabling facial recognition and building robots with personality and behavioral differences between the produced duplicates of particular robots.

In spite of the progress, there are still many challenges. These include making a robot that can conduct a comprehensive conversation with humans on a broad range of subjects, walking fast in a crowd without hitting anyone, and operating over an extended period of time without the need to recharge its batteries as well as using batteries and power harvesting capabilities that don't constitute a significant percentage of the robot weight.

Generally, to develop low-cost robots in large production volume there is a need for standard hardware and software platforms that will have the same kind of interchange-ability and compatibility as in personal computers. Thus, designers will not need to start from basic levels each time they develop a new model of robot. Moreover, there is a need to increase the speed of response of robots and their reaction to changes in their environment. These robots will need to be able to recognize significantly more words than they can today as well as perform better comprehension of verbal and text communication. The developed robots will require the use of many miniature light-weight sensors with distributed processing capability. Also, there is a need for effective actuators with high power density as well as high operation and reaction speeds. The use of networked wireless robots can benefit from the power of personal computers to handle complex tasks, including image and speech recognition, navigation, and collision avoidance. Future improvements will require more standard computer codes so that software integration can be done faster with minimal development effort on the part of the roboticists and at the manufacturing stages. Further, there is a need for advances in automated design and prototyping from macro scale locomotion to micro scale sensing, actuation, and drive electronics.

SUMMARY

Emulating the human appearance, abilities, and intelligence is the ultimate goal of efforts to reproduce human beings in art and technology. Producing humanlike robots is part of the broader effort to develop biologically inspired technologies – also known as biomimetics. Making humanlike robots involves engineering machines that copy human appearance and imitate our behavior as biological systems. Humanlike robots, which were once considered science fiction, are today walking into our lives, thanks to recent advances in technology. Tools such as finite element modeling, computer simulations, speech recognition, rapid image processing, graphic displays, and animated simulations, as well as many others are helping us to make enormous improvements in producing lifelike robots.

There are still numerous challenges to producing humanlike robots. Addressing these challenges requires multidisciplinary expertise that includes electromechanical engineering, computational and material science, robotics, neuroscience, and biomechanics. The advances in humanlike robots are greatly supported by progress in many related fields, including AI, artificial muscles, artificial vision, speech synthesizers, mobility, and control. A graphic illustration of a humanlike robot posing a challenge to its audience is shown in Figure 8.5, which highlights the potential of this technology.

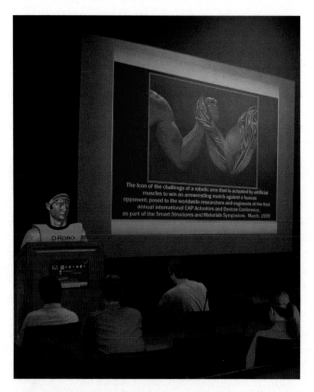

Figure 8.5. A humanlike robot shows graphics of the arm wrestling challenge using robots driven by artificial muscles. This image highlights the challenges of this technology and its great potentials.

As these robots become more useful in our lives they may start appearing as household appliances or our peers and probably will become a common sight in our future environment. One could envision a friendly, intelligent robot being welcomed in future homes. A visionary graphic illustration of such a friendly, welcomed robot is shown in Figure 8.6.

There are three key areas with great challenges to be met. These are as follows:

- *Tele-present humanlike robots.* Such a robot is remotely controlled by a human. The human has a suit with sensors that are linked in a mirrored fashion to the robot, and the robot simultaneously follows the movements of the human operator. The human is equipped with three-dimensional goggle display to allow immersing himself or herself in the environment of the robot to be able to operate and react as necessary. To feel the conditions that are encountered on the robot side there is a need for effective haptic interfacing that allows sensing and projecting to the user the forces, temperatures, and other physical conditions. An example of such a tele-presence control is done with NASA's Robonaut (see Chapters 1 and 2).
- *Cyborgs using a natural human brain with the rest of the body in a machine form.* Such a condition may occur if a human who has a terminal body condition but a healthy brain seeks to continue to live. In such case, a mechanized body may be the only choice. For this purpose, a bio-electro-mechanical body that is humanlike would be developed to respond to the brain's control commands, and it would be designed to keep the brain alive by providing the necessary physiological requirements such as maintaining proper temperature, fluids, oxygen, blood circulation, and others. We are not even close to having this available.
- *Fully autonomous humanlike robots.* Making such a robot behave identically as a human is a great challenge and will probably take many years before it becomes a reality. The possibility and consequences of instilling cognition into a robot was discussed throughout this book and particularly in Chapter 7. Although it is far from a near future realization, this goal is expected to be reached gradually as technology advances.

Figure 8.6. A futuristic illustration of a friendly intelligent robot assistant. The photo of Yoseph Bar-Cohen (*right*) was taken by Yardena Bar-Cohen and was modified by Adi Marom.

There are definitely strong societal differences in attitude toward the technology of humanlike robots in the West and in the Pacific Basin (mostly Japan, Korea, and China). Besides fundamental differences in religious beliefs about the nature of inanimate objects such as robots, many Japanese scientists grew up as admirers of robots. From childhood they were presented with favorable robotic cartoon characters and films, such as the animation movie *Mighty Atom,* whose main character tries to keep peace among humans. Also, as opposed to the United States and Europe, Japan is in a unique position in that it does not want to import foreign laborers to perform such tasks as take care of its elderly. So it must rely on development of humanlike robots as an alternative. In contrast, in the West there are significantly more science fiction books and movies that present robots as evil doers that pose great dangers to humans. Also, in the West there is the feeling that such robots will have to be highly beneficial and provide great practical value before they could become a household appliance; it may not be sufficient that they are a novelty or cute, particularly if they are expensive.

Generally, movies have portrayed a frightening image of the possible consequences of having humanlike robots go out of control. On the other hand, some animated movies and cartoons are helping children to see such robots as a helpful part of our lives and possibly reducing fears of this technology. Children today view humanlike robots more as a reality than do adults. One of the authors of this book (Bar-Cohen) experienced this view when a group of high-achiever junior high-school students came to visit his lab at the Jet Propulsion Laboratory. At the end of their tour, one of the students looked at him directly in the eyes and asked, "Are you a robot?" Raising such a question would not have even crossed his mind in another time, but it is more of a reality for children today. This book shows that this possibility of humanlike robots operating in our offices, businesses, and homes may not be a dream anymore, and while we can expect great benefits we had better start preparing to see robots as products or even peers and deal with their potential impacts.

BIBLIOGRAPHY

Abdoullaev, A., *Artificial Superintelligence*, F.I.S. Intelligent Systems, (June 1999).

Bar-Cohen, Y. (Ed.), *Electroactive Polymer (EAP) Actuators as Artificial Muscles – Reality, Potential and Challenges*, 2nd Edition, SPIE Press, Bellingham, Washington, Vol. PM136, (March, 2004).

Bar-Cohen, Y., (Ed.), *Biomimetics – Biologically Inspired Technologies*, CRC Press, Boca Raton, FL, (November, 2005).

Bar-Cohen, Y., and C. Breazeal (Eds.), *Biologically-Inspired Intelligent Robots*, SPIE Press, Bellingham, Washington, Vol. PM122, (May, 2003).

Bartlett, P. N., and J. W. Gardner, *Electronic Noses: Principles and Applications*, Oxford University Press, Location, (1999).

Bialek, W., "Physical limits to sensation and perception," *Annual Review of Biophysics, Biophysics Chemistry*, Vol. 16, (1987), pp. 455–478.

Breazeal, C., *Designing Sociable Robots*. MIT Press, Cambridge, MA, (2002).

Craven, M. A., and J. W. Gardner, "Electronic noses – development and future prospects," Trends in Analytical Chemistry, Vol. 15, (1996), p. 486.

Dietz, P., *People Are the Same as Machines – Delusion and Reality of Artificial Intelligence,* Bühler & Heckel, German, (2003).

Hanson, D., "Converging the Capability of EAP Artificial Muscles and the Requirements of Bio-Inspired Robotics," *Proceedings of the SPIE EAP Actuators and Devices (EAPAD) Conference*, Y. Bar-Cohen (Ed.), Vol. 5385, SPIE, Bellingham, Washington, (2004), pp. 29–40.

Hecht-Nielsen, R., *Mechanization of Cognition*, in (Bar-Cohen 2005), pp. 57–128.

Hughes, H. C., *Sensory Exotica a World Beyond Human Experience*, MIT Press, Cambridge, MA, (1999).

Krantz-Ruckler, C., M. Stenberg, F. Winquist, and I. Lundstrom, "Electronic tongues for environmental monitoring based on sensor arrays and pattern recognition: a review," *Analytica Chimica Acta*, 426 (2001), p. 217.

Levy, D., *Love and Sex with Robots: The Evolution of Human-Robot Relationships,* Harper Collins Publishers, New York, NY, (2007a).

Levy, D., "Robot prostitutes as alternatives to human sex workers," *Proceedings of the IEEE-RAS International Conference on Robotics and Automation (ICRA 2007)*, Workshop on Roboethics, Rome, Italy, (April 10–14, 2007b).

Menzel, P., and F. D'Aluisio *Robo Sapiens: Evolution of a New Species*, The MIT Press, Boston, MA, (2000)

Nickel, B., *input was sent as an e-mail communication to Bar-Cohen*, (April 2, 2006).

Raibert, M. H., *Legged Robots that Balance*, MIT Press, (1986).

Russell, S. J., and P. Norvig, *Artificial Intelligence: A Modern Approach*, Pearson, New Jersey, (2003).

Szema, R., and L. P. Lee, *Biologically inspired Optical Systems*, in (Bar-Cohen, 2005), pp. 291–308.

Index

Printed in the United States of America